Honold    Secondary Radar

Fundamentals and
Instrumentation

# Secondary Radar

Fundamentals and Instrumentation

By Peter Honold

SIEMENS AKTIENGESELLSCHAFT
HEYDEN & SON LTD.

Heyden & Son Ltd, Spectrum House, Hillview Gardens, London NW4 2JQ
Heyden & Son Inc., 247 South 41st Street, Philadelphia, PA 19104, USA
Heyden & Son GmbH, Devesburgstrasse 6, 4440 Rheine, Germany

---

CIP-Kurztitelaufnahme der Deutschen Bibliothek

**Honold, Peter**

Secondary radar.–London, New York, Rheine/Westf.: Heyden;
Berlin, München: Siemens-Aktiengesellschaft, [Abt. Verl.], 1976.
Dt. Ausg. u.d.T.: Honold, Peter: Sekundär-Radar.
ISBN 3-8009-1226-0 (Siemens-AG)
ISBN 0-85501-248-X (Heyden)

---

Title of German original edition:
Sekundär-Radar
Von Peter Honold
Siemens Aktiengesellschaft 1971
ISBN 3-8009-1089-6

ISBN 3-8009-1226-0 Siemens AG, Berlin and München
ISBN 0-85501-248-X Heyden & Son Ltd, London

Reprinted 1981
© 1976 1981 by Siemens Aktiengesellschaft, Berlin and München; Heyden & Son Ltd., London.

All Rights Reserved. No part of this publication may be reproduced, stored in a retrieval system, or transmitted, in any form or by any means electronic, mechanical, photocopying, recording or otherwise, without the prior permission of Siemens Aktiengesellschaft and Heyden & Son Ltd.

Printed in Great Britain by The Whitefriars Press Ltd., London and Tonbridge

# Preface

In the last few decades radar techniques have been developing at a very rapid pace. Because of advances in engineering technology we are now in a position to make substantial increases in the performance of our equipment, whilst quite often reducing overall size and weight.

By improved methods of analysing the echo signals, information – other than the original basic requirements of range and bearing – can be obtained. For example, the ability to discriminate between fixed and moving targets by measuring the 'Doppler' displacement between the frequency of the transmitted signal and that of the echo.

Despite this and other successes, these techniques alone did not positively identify aircraft, with the result that, in the Air Traffic Control environment, radar could not be fully exploited. This problem was realized when radar was first introduced into military service, when upon locating a target it was necessary to decide whether it represented a friendly or enemy aircraft. In the last war therefore[1-3] both sides introduced equipment to supplement primary radar to assist with identification. This supplementary equipment became known as IFF (Identification Friend or Foe) and equipment used for military applications is still known as such. Similar equipment derived from war time IFF and used by Civil Aviation is known as SSR (Secondary Surveillance Radar).

The original primary radar, therefore, only revealed its total potential when associated with SSR so that now Air Traffic Control systems can rely on the integrity of the information provided.

The author has been working for many years on the development of secondary radar equipment, and throughout this period he has introduced many colleagues to this subject. Through lectures he has also familiarized many users of SSR installations with the techniques used. Some of these lectures were recorded in manuscript form and the wish was eventually expressed that the knowledge acquired by the author and his colleagues over the last 10 years should be put at the disposal of a larger circle of readers in a compact form.

An effort has been made to arrange the contents of the book in such a way that non-specialist engineers, unacquainted with telecommunications techniques, may obtain a basic understanding of this branch of engineering. A further purpose has been to provide a useful handbook for all those engineers, technicians, and operating staff who are actually responsible for the development, construction, maintenance, and operation of such equipment and installations. With such aims in view, it is inevitable that, for some readers, a few sections will seem to be unnecessarily explicit. However, it is the author's belief, based on experience and frequent discussions, that if non-specialists are to become properly acquainted with the system, it is precisely these basic principles that must be presented in such great detail.

An attempt has been made to introduce readers to the specialist works on the subject via a limited number of references, carefully chosen in relation to the scope of the book.

München, July 1976

SIEMENS AKTIENGESELLSCHAFT

# Contents

| | | |
|---|---|---|
| Preface | . . . . . . . . . . . . . . . | 5 |
| **1.** | **Basic Principles** . . . . . . . . . . . | 11 |
| **1.1.** | **Relationship Between the Problem to be Solved and the Characteristics of a Secondary Surveillance Radar System (SSR)** . . . . . . . . . . . . . . . | 14 |
| **1.2.** | **Coding for an SSR Secondary Radar System** . . . . | 21 |
| 1.2.1. | Encoding the interrogation . . . . . . . . . | 21 |
| 1.2.2. | Encoding the response . . . . . . . . . . | 22 |
| 1.2.3. | Transmission of altitude . . . . . . . . . | 24 |
| **1.3.** | **Evaluation and Presentation of the Response Information** | 29 |
| 1.3.1. | Raw video indication . . . . . . . . . . . | 29 |
| 1.3.2. | Passive decoding . . . . . . . . . . . | 31 |
| 1.3.3. | Active decoding . . . . . . . . . . . . | 32 |
| 1.3.4. | Automatic decoding . . . . . . . . . . . | 33 |
| **1.4.** | **Combination With the Primary Radar Unit** . . . . . | 34 |
| **1.5.** | **The Problems of Antennae** . . . . . . . . . | 35 |
| 1.5.1. | Omni-directional antennae . . . . . . . . . | 36 |
| 1.5.2. | Directional antennae . . . . . . . . . . | 38 |
| 1.5.3. | Matching and polarization . . . . . . . . . | 41 |
| 1.5.4. | Coordinating the interrogator antenna with the primary radar antenna . . . . . . . . . . . . . | 43 |
| **1.6.** | **Side-lobe Suppression (SLS)** . . . . . . . . . | 46 |
| 1.6.1. | Interrogation path side-lobe suppression (ISLS) . . . . | 47 |
| 1.6.2. | Reply-path side-lobe suppression (RSLS) . . . . . . | 51 |
| **1.7.** | **Fruit and Defruiters** . . . . . . . . . . . | 52 |
| **1.8.** | **Garbling** . . . . . . . . . . . . . . | 59 |
| **1.9.** | **Round Reliability of the System** . . . . . . . . | 60 |
| 1.9.1. | The transmission channel . . . . . . . . . | 60 |
| 1.9.2. | The transponder . . . . . . . . . . . | 65 |
| 1.9.3. | The defruiter . . . . . . . . . . . . | 67 |
| 1.9.4. | Decoding the response . . . . . . . . . . | 67 |

| | | |
|---|---|---|
| 1.10. | Possible Development in Future SSR Systems | 69 |
| 1.10.1. | An address-coded SSR system – selective interrogation | 70 |
| 1.10.2. | The data-link capability | 74 |
| 1.10.3. | The Synchro-DABS method | 74 |
| 1.10.4. | Airport surface surveillance | 75 |
| | | |
| **2.** | **Design of Interrogator Units** | **77** |
| 2.1. | Specifications for the Transmitter and Receiver Units in Secondary Radar Interrogation Units | 84 |
| 2.2. | Concepts of Signal Evaluation | 87 |
| 2.2.1. | The common decoder | 87 |
| 2.2.2. | Automatic decoders | 89 |
| 2.3. | Operational Reliability and Monitoring | 90 |
| 2.3.1. | Civil air traffic control units | 90 |
| 2.3.2. | Military units | 91 |
| 2.4. | Secondary Radar Interrogator Unit Type 1990 | 93 |
| 2.4.1. | The coder | 96 |
| 2.4.2. | The transmitter | 98 |
| 2.4.3. | SLS switching unit | 100 |
| 2.4.4. | The receiver | 103 |
| 2.4.5. | The logarithmic IF amplifier and the processing of the signal | 104 |
| 2.4.6. | The defruiter | 112 |
| 2.4.7. | The decoder tray | 114 |
| 2.4.8. | The multi-channel decoder | 121 |
| 2.4.9. | Remoting equipment | 126 |
| 2.4.10. | Characteristics of the secondary radar interrogation unit, Type 1990 | 128 |
| | | |
| **3.** | **The Secondary Radar Transponder** | **132** |
| 3.1. | Specifications for Secondary Radar Transponders | 133 |
| 3.2. | IFF Transponder AN/APX-46 | 139 |
| 3.2.1. | Design and block diagram | 139 |
| 3.2.2. | The RF unit | 143 |
| 3.2.3. | The IF amplifier unit | 145 |
| 3.2.4. | The decoder unit | 148 |
| 3.2.5. | Reference signal generator | 150 |
| 3.2.6. | The coder unit | 151 |
| 3.2.7. | The transmitter unit | 153 |

| | | |
|---|---|---|
| 3.2.8. | Power pack | 155 |
| 3.2.9. | Test unit | 155 |
| 3.2.10. | Control unit C4083/APX | 157 |
| **3.3.** | **Testing and Monitoring Unit T29** | 158 |
| **3.4.** | **Recent IFF Transponders** | 163 |
| 3.4.1. | IFF transponder AN/APX-90 | 163 |
| 3.4.2. | IFF transponder STR 700 | 166 |
| **3.5.** | **SSR Transponder for Air Traffic Control (ATC)** | 173 |
| 3.5.1. | ATC transponder AVQ-65 | 173 |
| 3.5.2. | ATC transponder SSR 2100 | 174 |
| 3.5.3. | ATC transponder 506A | 176 |
| 3.5.4. | ATC transponder UAT-1 | 179 |
| 3.5.5. | ATC transponder AT 6-A | 183 |
| 3.5.6. | ATC transponder KT 75 | 187 |
| 3.5.7. | ATC transponder TPR-610 | 188 |
| **4.** | **Various Applications of the Secondary Radar Process** | 191 |
| **4.1.** | **In the Military Sphere** | 191 |
| **4.2.** | **Civil Air Traffic Control** | 191 |
| **5.** | **Some Technical Expressions used in Secondary Radar Techniques** | 196 |
| **References** | | 213 |
| **Index** | | 214 |

# 1. Basic Principles

Secondary radar is a radio location system which measures time but which, in contrast to normal radar techniques (Fig. 1.1), instead of using the passive echo reflected from a target, uses an active answering device (the transponder) which is located in the target aircraft. In this system the ground station, the interrogator, effectively asks a question which, one might say, is 'answered' by the transponder located in the aircraft.

Although a secondary radar system will obviously give a position in terms of range and bearing, it is usually used in conjunction with primary radar. The reason for this is obvious: namely, that a secondary radar system requires cooperation, and assumes that a transponder is available. This, however, cannot at present be guaranteed, particularly in the case of general aviation aircraft.

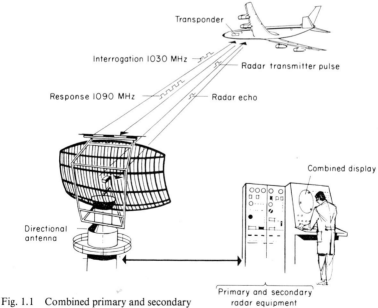

Fig. 1.1 Combined primary and secondary radar installation

In comparison with the primary radar system, the introduction of a transponder provides substantial advantages, viz.

(a) In contrast with the primary radar system where, as the range, $R$, increases the power of the echo decreases by a factor of $1/R^4$, the power of the transponder reply only decreases by a factor of $1/R^2$. Consequently, it is possible to work with a much lower transmitter power. To obtain some idea of this difference, assume that a peak pulse power of 1.5 kW is sufficient for a secondary radar installation to cover an area having a radius of 370 km. In a typical primary radar installation operating in the L band, a peak pulse power of some 1.5 MW would be required to cover the same range.

(b) The interrogation and the response can be transmitted at two different frequencies, thus avoiding any undesirable echoes; for example, ground clutter, permanent echoes, and such echoes as are caused by rain clouds and other meteorological phenomena.

(c) The transponder equipment is a receiver and transmitter of coded messages. Thus, an exchange of information can be obtained as well as information about location.

The use of a cooperative system assumes some standardization. For a Secondary Surveillance Radar system (SSR), also known in the U.S.A. as an Air Traffic Control Radar Beacon System (ATCRBS), these standards have been established by the International Civil Aviation Organization (ICAO) which is a subsidiary of the United Nations Organization (UNO). One of the essential tasks of ICAO is to coordinate all those problems of air traffic and air traffic control which require some international agreement. One of these tasks is to specify international standards for equipment and to recommend operational procedures. These technical specifications are published in the form of an appendix (Annex) to the ICAO convention, and must be observed by all member states.

Supervision of the secondary radar system is controlled by the Communications Division of ICAO. After lengthy preliminary discussions this division, at its Fifth Conference in Montreal in 1954, specified the frequencies to be used and other characteristics as an essential prelude to a trial run.

At its Sixth Conference, in 1957, also in Montreal, agreement was reached on almost all the characteristics which are mandatory today for the SSR system.

Further details, particularly those concerned with side-lobe suppression and altitude coding, were appended to Chapter 8 at the Seventh Conference of the Communications Division. All the ICAO regulations on the subject of SSR have, since then, been available in Annex 10 to the Convention for International Civil Aviation.[4]

There is also a military secondary radar system, known as IFF Mark X-SIF (IFF, Identification Friend or Foe; X, currently No. 10; SIF, Selective Identification Feature).

The electrical characteristics of the civil SSR system are substantially the same as those of the military identification system IFF Mark X-SIF. This considerably simplifies the important cooperation between the military and civil air traffic control authorities. To ensure this cooperation, further national regulations have been issued; for example, those in force in the U.S.A., the U.S. National Standard for Common System Component characteristics for the IFF Mark X-SIF/Air Traffic Control Radar Beacon System (ATCRBS) are published there by the Federal Aviation Authority (FAA).

Reference must be made to the fact that the applications of the secondary radar principle are not solely confined to the problems of air traffic control. The advantages of the secondary radar technique can be used in a number of other applications.

For example, when tracking the trajectories of rockets, the sole task of a secondary radar system is often merely to increase the echo power for the radar equipment. No transmission of elaborate information is required. Moreover, in this application, the answer can be obtained at the radar frequency, outside the radar bandwidth, or at a frequency with a constant shift relative to the radar frequency. In English usage one then speaks of In-Band-Beacon, Out-Band-Beacon and Frequency Shift Beacon.

Such beacon systems are mainly used for very special applications, for which only limited numbers of special interrogators and transponders are required.

Another application of the secondary radar principle is to be found in the range-measuring section of the TACAN navigation system (Tactical Air Navigation System), and the civil DME system (Distance Measuring Equipment) which operate on similar principles.[5] In the two navigational systems the interrogator equipment is mounted in the aircraft, and the transponders are erected on the ground at marker beacons. The range measured is that from the aircraft to the beacon.

Although the two applications mentioned here are clearly part of our subject, their treatment is beyond the scope of this book. The following text will therefore deal exclusively with the problems of secondary surveillance radar systems (SSR).

## 1.1. Relationship between the Problem to be Solved and the Characteristics of a Secondary Surveillance Radar System (SSR)

The SSR-system is so designed that a ground station can monitor an air space having a maximum radius of 200 nautical miles, i.e. 370 km (1 nautical mile = 1.85 km) and a height of some 15 km above the radar horizon. In this space only a few aircraft, relative to the size of the area, and separated at great distances from one another, will be moving at high speeds. In the radial direction the location of an aircraft must be accurate to within some 10 m and, in azimuth, must be accurate to within a few degrees so that these measurements can be correlated with the findings of the primary radar equipment. Using special codes, the identification information not only makes it possible to distinguish different aircraft but also facilitates the transmission of data such as altitude. Moreover, this air-monitoring system is so arranged that a large number of aircraft can be surveyed by a relatively small number of ground stations.

The following will explain, very briefly, how, from relatively simple technical considerations, the characteristics of the system can be derived merely by defining the problem − namely, location and the transmission of information. More precise data on the appropriate parts of the system will then be treated in detail in the following sections.

The first requirement is that all interrogator stations should be able to effect a dialogue with a number of transponders simultaneously, and that the range will be calculated from the time which elapses between the emission of the interrogation and the receipt of the response.

This results in an equal-channel system in which there is a fixed carrier frequency for every interrogation (1030 MHz) and a fixed carrier frequency for every response (1090 MHz).

Similarly, as in primary radar techniques, the type of modulation is pulse modulation. In the usual primary radar installations short, single, pulses are emitted at a fixed rate (in the order of magnitude of milliseconds), whilst in the secondary radar system information is exchanged in the form of groups of pulses which meet the required criteria for coding.

The following points are of critical importance when selecting the coding for the interrogations and responses. Due to the high speed of aircraft, the position 'picture' must be continually renewed at very short intervals. Consequently, the directional antenna of the radar installation usually rotates at a uniform speed, and usually scans the area to be covered at a rate of 5–15 rev min$^{-1}$ (Fig. 1.2) with a fan-shaped beam which is very narrow in azimuth (e.g. 1°) but is wider in elevation. With such a type of spatial exploration, each aircraft is only covered for a very short period of time with the result that, in a single beamwidth, some 10–30 interrogations – normally termed hits per scan – fall upon a target. During this short interval only a very small flow of information in any of the six interrogation modes can take place in the ground-to-aircraft direction. Expressing this more basically, a single call is given out 'to all'. The actual information is then only exchanged in the aircraft-to-ground direction. The separation of the information is effected upon receipt of the responses. It is therefore essential, as a result of the spatial distribution of the targets, that the responses should generally be received in a temporal sequence. For instance, if two or more targets are simultaneously within the scanning range of the directional interrogating antenna, and the difference between the distances separating them from the interrogating unit is less than the range equivalent to the time required for a response request, then, on reaching the ground station, they will interfere with one another and may sometimes cause mutual falsification of their messages. To avoid such confusion, usually known as garbling, the response pulse train is made as brief as possible (20.3 μs).

The next problem concerns the frequency of the interrogations. The upper limit for this interrogation recurrence frequency (IRF) is determined by the fact that the response from the most-distant target must be received before

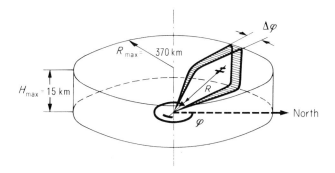

Fig. 1.2  Scanning the airspace with a fan-shaped beam

Defined for a target with an azimuth angle $\varphi$ (relative to north) and a slant range $R$

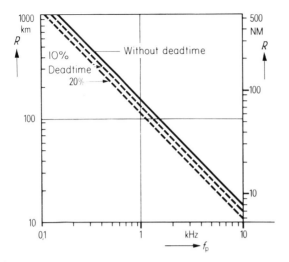

Fig. 1.3 Unambiguous range $R$ as a function of the radar pulse repetition frequency $f_p$

*Note:*
In actual practice it is not possible to use the maximum pulse repetition frequency for a given range. For technical reasons a pause, known as the *dead time*, is usually inserted between the receipt of the echo from the maximum range and the emission of the next transmitter pulse. The *dead time* usually occupies 10–30% of the total duration of the period

transmitting a fresh interrogation, to ensure that the range measurement can be made without any ambiguity (Fig. 1.3). In the instance of the maximum required range – namely 370 km – this means a signal period of 2.5 ms to cover the outward and return journey. Consequently, to cover the full range, an interrogation repetition-frequency of less than 400 Hz would be selected (the maxium IRF is fixed at 450 Hz by the ICAO).

In addition to this requirement for an unequivocal range measurement, precautions must also be taken to ensure that the results of the interrogation can be correlated with the primary radar signals. For this reason the secondary radar equipment operates either at the same pulse repetition frequency as the primary radar equipment, or at a sub-harmonic of it.

After what has already been explained, one still has to answer the question: how is it possible that so many interrogator and transponder equipments can operate simultaneously at the same interrogation and response frequencies without any interference? The maximum interrogation recurrence frequency (IRF) is 450 Hz, i.e. the interval separating two interrogations must be at least 2200 μs. The actual message only takes 20.3 μs of this period, so that the remaining period of 2180 μs is available to other aircraft. Moreover, the interrogation is effected by a rotating

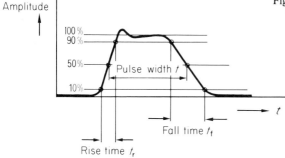
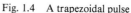

Fig. 1.4 A trapezoidal pulse

directional antenna so that the transponder only answers during the brief interval of time within which the fan-shaped beam scans the target. For the normal breadth of the lobe, which is 3.6°, this is but 1/100 of the total time. Thus, with a completely uniform distribution, some 10 000 transponder equipments could operate at the same time. What happens when the distribution is not uniform will be discussed in a later section.

One is still left with the problem of the shape of the pulse in the response (Fig. 1.4). The pulse width $t$ has been fixed at $0.45 \pm 0.1$ μs as a result of the requirement that the reply-pulse must have a maximum of $2^{12} = 4096$ code words within two, fixed, framing pulses, and an additional, empty, position in the centre of the pulse.

These 15 pulse positions must be contained within the period of 20.3 μs already mentioned. They must also occur in such a way that they give a mark:space ratio of approximately 1:2. By this means the pulses can be fairly certainly recognized, even if they become distorted during their transmission.

The individual pulses, however, have a further criterion, i.e. their rise and fall time, which is measured respectively at the leading and trailing edges, between 10% and 90% of the maximum amplitude. As is well known from Fourier analysis, the steeper the sides of the pulse become, the broader becomes the frequency spectrum, and, consequently, the bandwidth, that has to be transmitted.

For this reason it is necessary to decide what pulse rise-time is absolutely essential. As was mentioned earlier, a determination of range in the secondary radar process is also derived by measuring the period of time between the interrogation and the receipt of the response.

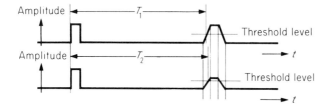

Fig. 1.5 Error in time measurement of two pulses of different amplitude when using the same threshold level for measurement

To be able to measure, automatically, the position of any pulse with respect to time, it is essential to determine at what moment the instantaneous value of a voltage pulse exceeds a specified reference level. All the responses, however, are not received at the same amplitude since, even at the same range, considerable fluctuations in amplitude must be expected in the transmission channel. It is therefore obvious that, despite an identical position in time, a pulse of lower amplitude will attain the reference level later than a pulse of higher amplitude (Fig. 1.5).

As a consequence, the time measurement will be slightly greater for the smaller pulse and will therefore indicate an apparently greater range. It is, of course, possible to control the reference level of the maximum amplitude in such a way that these sources of error can largely be eliminated. In actual fact, the signal pulse is never *ideally* transmitted but, usually, has some degree of noise superimposed upon it. Hence, even with a sliding reference level, an error in the time measurement can occur depending upon the situation of the components of the signal (Fig. 1.6).

The steepness of the leading edge is determined by the fact that as regards its monitoring function, the range measurement may be subject to an error

Fig. 1.6 Error in measuring the time between two pulses due to the superposition of an interference voltage

of 15 m, which corresponds to a maximum rise time of 0.1 µs. The fall time is not directly used for the measurement of range; the maximum time specified for this is 0.2 µs. The spectrum for such a pulse 'train' is covered by a video bandwidth of approximately 5 MHz, a very common value for the usual television video signal.

After discussing the response pulse train, some mention must be made of the interrogation signal. Since the variety of coded messages is much smaller (six Interrogation Modes) the signal, which can be much more simply produced, consists of two pulses which are coded by means of the separation between them. For the interrogation pulses, the pulse width is $0.8 \pm 0.1$ µs.

The following considerations determine the choice of the carrier frequency.

Frequencies are distributed in accordance with an international scheme, according to which specific frequency bands are allotted to various radio services, so that mutual interference is thus avoided. Such a distribution of bands is also provided for radar equipments. Depending on their purpose and type these bands may be of quite different frequency ranges. Thus, long-range radar equipments make use of long wavelengths, whilst very short wavelengths are used for very short-range installations with high resolution. It has become usual to distinguish the separate bands for radar frequencies by an alphabetical letter code (Table 1), which gives a rough estimation of the operating frequency.

If cooperation with radar installations is to be possible, the frequency for the SSR-system must also lie within one of these bands.

Table 1  Alphabetical letter code for radar frequency bands

| Band | Approximate frequency (GHz) | Approximate wavelength (cm) |
| --- | --- | --- |
| P | 0.3 | 100 |
| L | 1 | 30 |
| S | 3 | 10 |
| C | 6 | 5 |
| X | 10 | 3 |
| K | 20 | 1.5 |
| Q | 37.5 | 0.8 |

The separate frequency-ranges within these bands are more accurately distinguished by indices, e.g. $K_u$-band (15.35–17.25 GHz). These notations are not, however, standardized and are not always uniformly used in the relevant literature.

By virtue of the attenuating properties of the Earth's atmosphere, the range required by the system can only be attained in the long wavelength regions. Moreover, a special section is also reserved in the L-band for radio navigation purposes. The interrogation frequency of 1030 MHz, and the response frequency of 1090 MHz, have been set within this range. The spacing between the interrogation and response frequencies is therefore 60 MHz, which is sufficient to keep the interrogations and responses separate from one another in equipments with (technically simple) separating filters.

The transmitter output power, and the receiver sensitivity, are also to be deduced from the required range. If the technical possibilities and the differing number of ground and aircraft equipments are to be taken into consideration, these values must be different for the ground- and aircraft-mounted installations. In a ground installation the peak pulse power is 1.5 kW, whilst for the aircraft installation it is 500 W. To compensate for this, the receiver sensitivity on the ground equipment is approximately $-82$ dBm, whilst on the aircraft installation it is some $-75$ dBm. These characteristics are compiled for clarity in Table 2.

Table 2  Electrical characteristics of SSR and Mark X-SIF systems

| | |
|---|---|
| Range of system | $R =$ 200 nautical miles / 370 km |
| Frequency band | L-band |
| [a] Interrogator frequency | $f_1 = 1030$ MHz |
| corresponding to a wavelength | $\lambda_1 =$ 29 cm |
| [a] Response frequency | $f_2 = 1090$ MHz |
| corresponding to a wavelength | $\lambda_2 =$ 27.5 cm |
| [a] Transmitter power (peak pulse power) | |
| Ground installation (Interrogator) | $P_{t_1} = 1500$ W |
| Aircraft installation (Transponder) | $P_{t_2} = 500$ W |
| [a] Receiver sensitivity (sufficient for proper decoding) | |
| Ground installation (Interrogator) | $P_{r_1} = -82$ dBm |
| Aircraft installation (Transponder) | $P_{r_2} = -75$ dBm |
| Pulse rise time | $\tau \leqslant 100$ ns |
| Video bandwidth | $B_{vid} = 5$ MHz (approx.) |
| Antenna for interrogator installation | Directional antenna |
| Antenna for transponder installation | Omni-directional antenna |
| Polarization | Vertical |

[a] Suffixes 1 and 2 refer to interrogator and transponder respectively.

## 1.2. Coding for an SSR Secondary Radar System

The general principles which determine the selection of the characteristics have been indicated in Section 1.1, so we will now discuss the actual coding of the interrogation and the response.

It must be noted that various terms, which in no way agree with the terminology of coding experts, have become common usage in the practical treatment of secondary radar processes, and have since been accepted in the ICAO regulations. Thus, the code word for interrogations is known as the 'interrogation mode', so as to provide a distinct difference, in speech, from the 'response code'.

However, the code word for the response is simply named, for brevity, 'the code'.

These terms, *interrogation mode* as the code word for the interrogation, and *code* as the code word for the response undoubtedly have the advantage of brevity, and the avoidance of confusion. For this reason they will be used throughout this book.

### 1.2.1. Encoding the interrogation

It has already been shown that, as a means of interrogation, only one 'call to all aircraft' is practically possible. The actual distinction between targets can only be made by means of the response code. However, six different types of interrogation mode can be encoded (Fig. 1.7). With these six different interrogation modes it is possible to ask all aircraft equipped with a transponder six different questions.

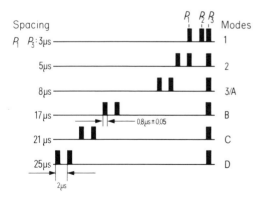

Fig. 1.7 Encoding the interrogations for the different interrogation modes

Table 3  Synopsis of the six interrogation modes and their application

| Mode | Pulse spacing (μs) | Application |
|---|---|---|
| 1 | 3 | Military IFF |
| 2 | 5 | Military individual code |
| 3/A | 8 | Military IFF / Military air traffic control / Civil air traffic control |
| B | 17 | Civil air traffic control |
| C | 21 | Air traffic control/altitude transmission |
| D | 25 | Civil air traffic control/not yet in use |

Besides interrogating in one single mode it is also possible, depending on the operational problem, to connect together several interrogation modes in a continuous sequence.

A commonly used method of interrogation mode interlace asks, alternately, for identification and altitude; i.e. the interrogations follow one another alternately according to the pattern A C A C A C.

The interrogation mode is defined by the spacing between two pulses which are denoted $P_1$ and $P_3$. The significance of the extra pulse, $P_2$, which follows 2 μs after $P_1$, will be explained later.

Table 3 gives a synopsis of the characteristics of the six interrogation modes and their applications. It should be noted that, in mode 3/A, the civil and military systems overlap one another thus permitting it to be used by both systems.

### 1.2.2. Encoding the response

In an SSR secondary radar system the targets, as is well known, are distinguished by the type of response. A pulse train (Fig. 1.8) will serve this purpose if it contains two framing pulses, $F_1$ and $F_2$, separated by an interval of 20.3 μs and can take up to 12 information pulses, in fixed positions, within the intermediate 1.45 μs time scan. The centre pulse, appearing (with respect to time) at the 10.15 μs position and known as the X pulse, is always 'empty'. From the presence or absence of these 12

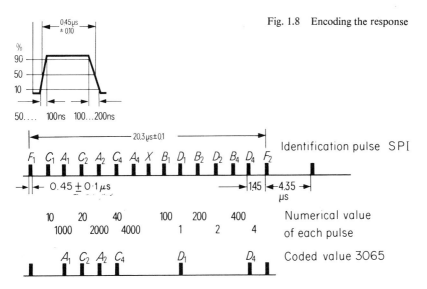

Fig. 1.8 Encoding the response

separate information pulses it is possible to form $2^{12}=4096$ different response code words. To simplify recognition and coding the information pulses are divided into four groups of three (A, B, C and D), and are encoded in binary, octal, form. Within each group of three, each pulse has a specific numerical value (significance), e.g. $A_1=1000$, $A_2=2000$, and $A_4=4000$. The code thus appears as a four-digit number (ignoring the figures 8 and 9).

A further pulse, denoted by SPI (Special Position Identification), can follow the response telegram at a distance of three scan positions, i.e. 4.35 µs after the second framing pulse. This pulse can be added to any code word to give an identification of position (I/P).

Earlier transponders, which are still in use, only have six positions for information, which reduces their coding facilities to $2^6=64$ code words.

When digital data is transmitted over a radio channel there is always the danger that some of the information may be lost through some interruption in its propagation. On the other hand, pulses can be simulated when there is radio interference with the result that the information will also be seen to be incorrect.

Of course, in data transmission techniques there are innumerable processes whereby the actual information is provided with supplementary safety devices such that the receiver location can determine whether the message has been correctly transmitted.

Secondary radar processes do not make use of such methods for securing the integrity of the transmitted data. All 12 pulse positions are used for the transmission of information. This is made possible by the usual method of scanning, whereby many messages can occur whilst the directional aerial is scanning the target with the result that the same information is transmitted several times.

By means of this multiple transmission, the sequential information can be compared during the decoding process and the integrity of the data can therefore be checked.

It must, however, be noted that, in fact, the response code only represents a single number. The actual significance of this number can only be decided in conjunction with the triggering interrogation mode.

### 1.2.3. Transmission of altitude

In addition to the transmission of pure code numbers the transmission of altitude information in secondary radar engineering is a matter of particular importance. It is the answer to an interrogation in mode C (21 µs pulse spacing). The secondary radar system is here used, in the first instance, not merely as an identification system but is also transmitting navigation data from the aircraft to the ground. There is also a possibility that, before very long, some agreement will be reached regarding the application of the interrogation mode D which is not as yet in use – perhaps for transmitting airspeed, flight direction or some such data.

Some brief details regarding air traffic control are given to indicate the vital importance of this transmission of altitude.

When applying instrument flight rules (IFR) *en route* air traffic is controlled in so-called flight lanes which lie between two radar beacons. The breadth of these flight lanes is usually 18 km, to take account of the navigational accuracy of the aircraft. When aircraft are flying at the same height, in the same lane, to accommodate navigational accuracy and to avoid the risk of collision, they must for the sake of safety be separated by a fairly long time interval. If however, it is necessary to have a fairly large number of aircraft simultaneously in one lane, they can be separated into different altitude layers. Such vertical staggering can be much closer together than any horizontal staggering – approximately 1000 ft between two aircraft flying in opposite directions – for the simple reason that the measurement of relative flying altitude can be made with great accuracy on board an aircraft. The barometric method is used for this purpose.[5] The air pressure is measured in the aircraft, and the altitude relative to sea-level, for an assumed sea-level pressure of 1013.25 mbar, is then calculated by use of the formula for pressure altitude

$$A = (18\,400 + 67 T_m) \log(p_0/p)$$

where $A$ is the altitude height above sea level (m); $T_m$ is the mean temperature of the intermediate column of air in degrees Celsius; $p_0$ is the air pressure at sea level; $p$ is the air pressure at the measuring position.

The mean air pressure $p_0$ at sea level is 1013.25 mbar, but can fluctuate by $\pm 5\%$ due to meteorological effects. At sea level this corresponds to an error of $\mp 400$ m of altitude. The temperature is also a variable, and is also a function of height.

For practical applications of the pressure altitude, the ICAO have defined a so-called *standard atmosphere* as well as several correcting processes to cover the instances where the *actual atmosphere* does not correspond with the standard atmosphere. For the standard atmosphere there is a calibration curve which gives air pressure as a function of the altitude above sea level. By use of a suitable computing mechanism the principal part of this calibration curve can be allowed for in the altimeter (between the pressure meter and indicator) in such a way as to give a linear scale. The deviation of the factually-available real atmosphere, from the *standard ICAO atmosphere* can be compensated for by means of a correcting scale on the altimeter.

The simplest type of altitude measurement is that made during a long-distance flight at high altitude such as, for example, over the Atlantic. The basis of such a measurement is simply the ICAO standard atmosphere since, due to the satisfactory safety distance from the ground, any error in the measurement of absolute height is trivial, and also because the relative error is the same for all aircraft, so that any vertical staggering is not affected.

The division of the flight lanes into different altitudes is thus a division into different layers of equal pressure which are denoted, on a scale, as so-called flight levels (FL) which correspond to 100 ft stages of the standard atmosphere. Thus, FL 240 indicates, for example, a flight altitude of 24 000 ft and, in the presence of a standard ICAO atmosphere, would correspond to an altitude of 24 000 ft above sea level.

Where the error in the measurement of absolute height has to be compensated, for example, before landing, the position is rather more difficult. The pilot does this by requesting, by radio, the airport control tower for the QNH value*, and setting this into the above-mentioned correcting scale on the altimeter. The QNH value is that value of air pressure (measured in millibars) which must be set into the calibration scale of the altimeter to ensure that, at the altitude of the airport (above sea level), the altimeter will indicate this altitude accurately.

A specified transition level (which is dependent on local and meteorological conditions) is also specified for each air traffic control area so that the transition from relative to corrected altitude can be unambiguously achieved.

Since it has now been demonstrated that from the point of view of navigation, it is practicable (and quite possible) to stagger aircraft at intervals of 1000 ft within flight lanes, the question now arises as to how this altitude can be measured from the ground with sufficient accuracy, approximately to within 100 ft, for the purposes of air traffic control.

Because the plan coordinates of an aircraft can be very accurately ascertained with a primary radar system, the obvious suggestion would also be to measure the altitude of an aircraft by means of a radar system. This method, however, is only successful at close ranges. It is used, for example, in precision approach radar (PAR) installation. If, however, it is necessary to determine (within the required accuracy of 30 m) the altitude of an aircraft flying a course at a distance of some 100 km, it will be necessary to determine the angle of elevation of the aircraft to within one minute of arc. This, however, makes essential a very high vertical aperture antenna which, moreover, must be located at such a height that reflections from the ground cannot falsify its polar diagram. Practical models of 3-D radar installations, i.e. installations which determine height as well as the plan coordinates, have as yet only been produced for military applications. Their estimations of height, for the reason already given, can only be very rough and are, consequently, not nearly accurate enough for air traffic control applications.

In this respect secondary radar techniques actually offer a very elegant alternative. The altimeter is provided in addition to the normal indicator scale, with a coded disc. On receipt of any query regarding altitude, this disc is sensed and the altitude information obtained is transmitted in the form of a response pulse train to the ground station (Fig. 1.9), where it can be decoded into a numerical form and be assigned to a target aircraft as a direct indication of altitude. To exclude operational errors during the

---

* In air traffic control an alphabetical code, with three or five positions, is used to transmit standardized information. The first letter is always Q (hence the name Q-Code).

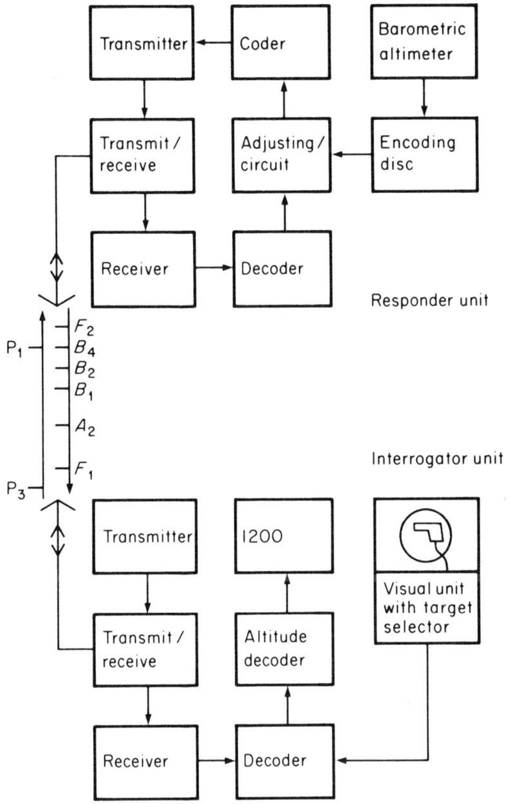

Fig. 1.9 Automatic transmission of altitude

transmission of altitudes, reference is solely made to the standard ICAO atmosphere with its pressure of 1013.25 mbar at sea level. If required, corrections will be made, during decoding, at the ground station.

To transmit altitude information the code used must satisfy the following requirements. The altitude information is an analogue value which is continually changing. During the change from one level of the digital code to the next, no serious error must occur through any imperfect sensing of the coding device. Possible inaccuracies must only result in the immediately higher or lower levels. These properties are, however, satisfied by the cyclic binary reflected code, a well-known type of this group of codes being the Gray code.

The same code must be suitable for the transmission of altitude, in steps of 100 ft or 500 ft. The reason for this is that very different types of aircraft

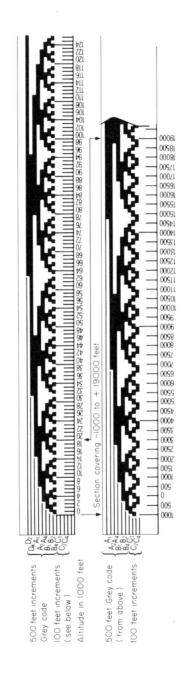

Fig. 1.10 Arrangement of the MoA-Gilham code

have to be equipped with instruments the accuracy of which must be economically appropriate. For example, at the lower end of the scale there is the private aircraft with a very simple transponder which only has a six-bit code; in this instance it is only possible to transmit altitudes between $-1000$ and $+30\,750$ ft, in steps of 500 ft. (Negative altitudes occur, especially at airports at sea level, before the QNH correction is applied.)

The next extension to a seven-bit transmission code is made possible by the addition of an SPI pulse, with the consequence that it is possible to transmit altitudes from $-1000$ ft to $+62\,750$ ft in 500 ft stages.

Transponders with as many as 12-bit codes are available for both civil and military aircraft. When using an 11-bit code it becomes possible to transmit altitudes, ranging from $-1000$ ft to $+126\,700$ ft, in 100 ft steps.

To obtain the required compatibility a special code, the MoA-Gilham code, has been created for altitude transmission. This code contains two parts. The first eight bits, which occupy the following pulse positions, $D_2$, $D_4$ (or SPI), $A_1$, $A_2$, $A_4$, $B_1$, $B_2$ and $B_4$, encode the altitude in 500 ft stages in a Gray code. By the addition of three further bits, in pulse positions $C_1$, $C_2$, and $C_4$, which represent another different cyclic binary code, fine coding in 100 ft stages is achieved. By itself, this fine coding is not without ambiguity. The removal of any ambiguity is only achieved in conjunction with the coarse code. The arrangement of the MoA-Gilham code is given in Fig. 1.10.

## 1.3. Evaluation and Presentation of the Response Information

As has already been explained, the received information is evaluated at the ground station. For this purpose the pulsed response signal must not only be coordinated, relative to position and time, with the primary radar signal, but it must also be produced in a form suitable for further processing. At present, four methods of processing are available.

### 1.3.1. Raw video indication

Raw video indication implies the direct presentation of the response without evaluating the contents of the message by simply focusing the raw video signal on the radar screen (Fig. 1.11a).

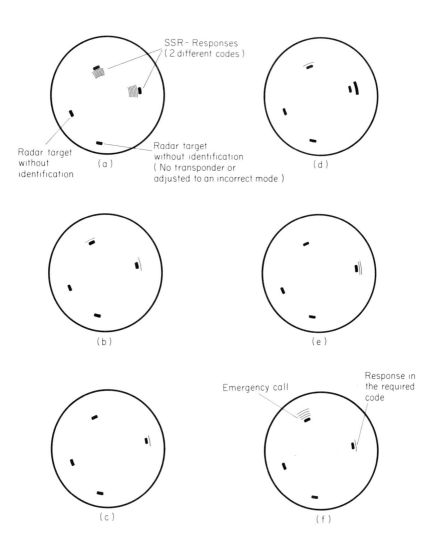

(a) Direct display of the response (Raw video)
(b) Identification of all aircraft which respond with the correct mode in the SSR system. All Common System (C/S)
(c) Identification of an aircraft responding with the required code in the correct mode
(d) Identification of all aircraft responding in the right mode with special preference for targets responding in the required response code ('beacon assist')
(e) Identification of an aircraft responding in the required code with which there is a direct radio conversation link (I/P)
(f) Identification of an aircraft in an emergency situation. This display is independent of all the other types of display.

Fig. 1.11   Displaying the response information when using passive decoding

*Note:*
The illustrations show the different types of display on the PPI radar screen for identical aircraft situations.

A raw video display only answers the question 'is the aircraft responding to the secondary radar's interrogation?'.

### 1.3.2. Passive decoding

Passive decoding implies the indication on the radar screen of an aircraft which is supplying the expected answer. Passive decoding thus answers the question 'Target X, where are you?'.

There are, however, several variations of this method of decoding, viz:

(a) Indicating all the aircraft which give an SIF response to the *interrogation mode* used (Fig. 1.11b).
(b) Indicating the aircraft which replies (with the correctly coded response), to the interrogation mode used (Fig. 1.11c).
(c) Indicating all aircraft which respond to the interrogation mode used, with special *preference* given to the aircraft which uses the correct response code (Fig. 1.11d). This method of indication is known as 'beacon assist'. It is a combination of the two previous types of display. Aircraft which merely answer the interrogation are indicated by a thin line, whilst those which respond with the correct code are indicated by a thick line.
(d) Indication of an aircraft in direct radio communication with the ground station whose position must be identified (Fig. 1.11e).

At the request of the ground station, the pilot switches the transponder to the I/P mode of operation (Identification of Position).

During the period of this conversation (and for 30 s thereafter) the transponder will either emit the response telegram twice, with an interval of 4.35 µs, or it will give the special SPI pulse.

The radar observer indicates the aircraft with which he is in contact by a double stroke on the radar screen. This mode of operation is of importance if several aircraft with the same identification signal appear simultaneously on the radar screen.
(e) To *indicate emergencies,* codes 7700 (general emergency) and 7600 (failure of radio communication) in mode 3/A have been agreed upon. For military units there is also an emergency call which involves repeating the response three times (Fig. 1.11f). The ground station contains special emergency-call decoding units which, on receiving an emergency call, immediately provide visual and audible alarms and, in addition, give a special display of the emergency call on the radar screen.

### 1.3.3. Active decoding

*Active decoding* implies the *numerical display* of the answer from any target selected on the radar screen. Active decoding answers the question 'Who are you?', or 'At what altitude are you flying?'.

The response signal relating to the target of interest is first selected and stored, after which it is then decoded and displayed. Typical installations are designed to display both types of information i.e. identification and altitude (Fig. 1.12). There are two different ways of selecting the target to be decoded.

The first uses a light pen which is manually placed over the required target, on the radar screen. The image thus formed, using an optical system and a photo-electric detector, is transformed into a trigger pulse for the active decoder.

In many radar installations the target can also be selected by a control stick, or a rolling ball, using an identification symbol or a marker gate. In this case the marker or symbol is overlaid on the target to be actively decoded.

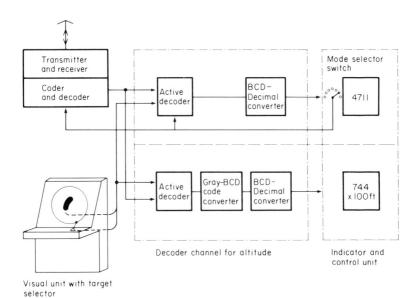

Fig. 1.12  Two channel active decoder

### 1.3.4. Automatic decoding

With the change to electronic data-processing, installations have also been developed for air traffic control where the data from the secondary radar units can also be stored in parallel with that from the primary radar units.

These intermediate units, between the recording of the data in the secondary radar unit and the processing of the data in the computer, are known in the U.S.A. as the Beacon Video Digitizer (BVD).

Such an automatic decoding device consists, basically, of the components shown in Fig. 1.13.

The standard video signal arriving from the secondary radar unit is received (after some of the noise has been removed), in real time, by an extractor or special type of computer operating at the same timing as the radar equipment, where it is subjected to initial evaluation, the filtering out of redundant information and, if necessary, of noise as well. The output of the extractor is connected to a buffer store. From this store another computing device – usually, the universally-employed electronic data-processor (EDP) – can extract the information, at its own timing, for further processing.

The functions of this EDP are:

(a) Correlation of the response information in all the different modes with the primary radar.

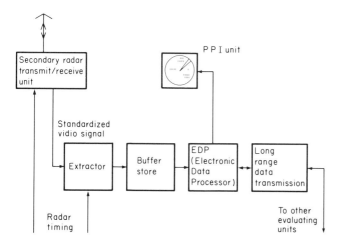

Fig. 1.13  Automatic decoding

(b) Validity of the data in an individual response can be checked by comparison with other responses received from the same aircraft in the same mode.
(c) Improving the location results by forming average values.
(d) Eliminating false location results from considerations of plausibility.
(e) The transformation of the coordinates of the location results.

As an intermediate result the following information, from all the aircraft which provide responses, is available in the data processor store: identification data in all the different modes, altitude information, x-coordinates, y-coordinates and quality classification of all the available information.

This information can either be transmitted on a narrow band link to a remote central store, or can be directly processed by the EDP, depending on the programming. One function will be the production of forecasts of aircraft position, and the automatic prediction of conflict situations.

## 1.4. Combination with the Primary Radar Unit

It has already been explained (Section 1.3.) that the information from the secondary radar is usually evaluated in conjunction with the positional information provided by the associated primary radar. To do this it is essential that the information from both the units should be capable of being unambiguously correlated and displayed.

*Angular correlation* is achieved by adjusting the directional patterns for the antennae of the primary and secondary radar units to overlay each other. The technical problems involved in these processes will be more thoroughly discussed in Sections 1.4 and 1.6.

To ensure an unambiguous *correlation of range* the secondary radar, as has already been mentioned, operates at either the same pulse-repetition rate as the primary radar unit, or, if this would cause any ambiguity in range, at a sub-harmonic frequency.*

Moreover, the secondary radar interrogations are emitted at a clearly specified interval before the emission of the associated primary radar pulse. This 'advance period' is so selected that, despite the duration in time of the following:

(a) The interrogation pulse (3 ... 25 μs)

---

* A sub-harmonic frequency implies a lower frequency obtained through dividing the radar frequency by some whole number.

(b) The pulsed response code train (20.3 μs)
(c) The processing time in the transponder (3 μs)
(d) The processing time in the decoder (up to 45 μs),

the evaluated answer will be available at the precise moment when the radar echo is received. This advance period is determined by the so-called pre-trigger, which the primary radar unit supplies to the secondary radar interrogator unit. The production of the pre-trigger is substantially dependent on the circuitry of the primary radar and, consequently, must be separately considered for each case. Difficulties can generally be expected with Moving Target Indicator (MTI) radar units (units which suppress fixed target displays), since there is a danger, when manipulating the radar synchronizer, that the other properties of the radar unit may be unfavourably affected.

To provide correct correlation of the information, the display unit must incorporate the following additional devices. During passive decoding the signals must be combined in a video mixer. If the pulse-repetition frequency of the primary radar is substantially greater than that of the secondary radar unit there may be the danger that the brightness of the secondary radar echoes will be lower than that of the primary radar echoes. The screen of a PPI has the property of integrating the illumination produced by a series of prf video pulses, so that the brilliance of the radar echo is partly dependent upon the prf in use. Where the prf of the secondary radar signal is lower than that of the primary radar, artificial SSR video pulses can be generated so that the same number of pulses for both appear on the display and hence both echoes have the same brilliance. A circulating store is used for this purpose, for example a delay line or a shift register.

## 1.5. The Problems of Antennae

A secondary radar system can be thought of as a communications link between interrogating and responder units, whereby the major part of the route comprises the propagation of electro-magnetic waves in free space. The antennae at each of the terminal positions thus produces the transition from waves, confined in a conductor, to free-space waves and vice versa.

These antennae must perform two functions; firstly, to ensure that the power generated is in fact emitted (matching) and, secondly, to ensure that the free-space wave moves in the direction required by the flow of information (directional pattern). In this context it is not of any consequence in which direction the transition between the free-space and guided wave occurs. As a result of the law of reciprocity, the antenna behaves in exactly the same way whether it is transmitting or receiving.

There is a very wide range of technical literature dealing with the general problems of antenna, but the special features of primary radar antennae are generally given adequate treatment in the works on radar.[6]

The following details can therefore be limited to a very general review, to a few handy formulae and to some typical problems associated with secondary radar systems.

Different requirements are made from the interrogating and responder units as regards the directional effects of antennae.

From the point of view of the *interrogator unit* the azimuth of the target must be measured as well as the range, if its location is to be determined. Use is made, for this purpose, of *directional antennae* with which the air space can be swept in accordance with the selected scanning method.

On the other hand, it is a requirement of the responder unit that it must always be prepared to receive and respond to interrogations from any direction. This can only be achieved with an *omni-directional antenna*.

### 1.5.1. Omni-directional antennae

The ideal radiation pattern for a responder unit is a spherical radiation pattern (an isotropic radiator). Such a radiator emits the same power, perfectly uniformly, in every direction in space (Fig. 1.14), but these ideal spherical radiators do not, of course, exist. A hypothetical radiator of this type however is generally accepted as a reference for the gain of an antenna.

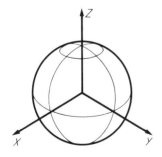

Fig. 1.14 Isotropic diagram for a (hypothetical) spherical radiator

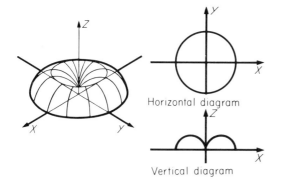

Fig. 1.15 Horizontal omni-directional radiation pattern for an λ/4 unipole on a conducting surface of infinite size

One of the limitations to the achievement of an omni-directional characteristic is the fact that, with the wavelengths of approximately 30 cm used in the SSR system, the aircraft fuselage screens some part of the total space.

λ/4 monopoles are, in practice, used as antennae for the responder units aboard aircraft. Figure 1.15 gives typical diagrams for one of these λ/4 monopoles, on a plane of infinite extent, in perspective view and in a horizontal and vertical cut.

An aircraft fuselage, however, is not an infinitely-extended surface but is a spatially-limited structure. This results in additional, considerable, deviations from the ideal form of diagram (Fig. 1.16).

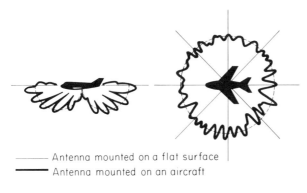

——— Antenna mounted on a flat surface
▬▬▬ Antenna mounted on an aircraft

Fig. 1.16 Antenna diagram for an SSR transponder antenna (λ/4-blade shaped antenna)

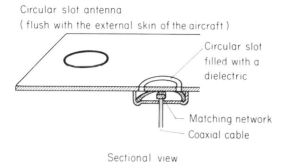

Fig. 1.17
Circular slot antenna

Because the antenna is fitted outside the aircraft, drag has to be considered. To reduce this, the antenna must have an aerodynamic shape, usually that of a blade (see Fig. 3.27), or a suitable covering.

These *blade-shaped antennae* actually have an excessive drag when fitted to supersonic aircraft. In such instances, therefore, frequent use is made of *flush-mounted annular slot antennae*. In an extremely simple form this type of antenna can also be considered as a degenerate $\lambda/4$ (quarter-wave) antenna, very much reduced in length due to capacitative loading, and set back slightly relative to the earth's plane (Fig. 1.17).

### 1.5.2. Directional antennae

There are, in principle, several different ways of producing a directional characteristic.

One method is to provide the wave field of a single emitter with an appropriately-shaped reflector (Fig. 1.19). In another method, the electromagnetic fields of a multitude of individual radiators can be superimposed to obtain the required directivity.

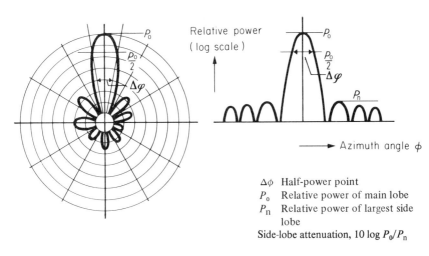

(a) Polar coordinates  (b) Cartesian coordinates

$\Delta\phi$  Half-power point
$P_0$  Relative power of main lobe
$P_n$  Relative power of largest side lobe
Side-lobe attenuation, $10 \log P_0/P_n$

Fig. 1.18  Graphical representation of directional patterns

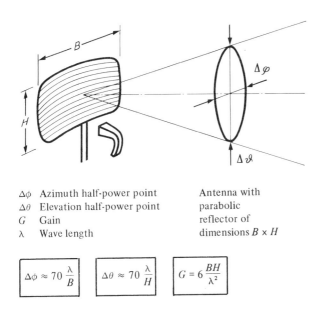

$\Delta\phi$  Azimuth half-power point
$\Delta\theta$  Elevation half-power point
$G$  Gain
$\lambda$  Wave length

Antenna with parabolic reflector of dimensions $B \times H$

$$\Delta\phi \approx 70 \frac{\lambda}{B}$$  $$\Delta\theta \approx 70 \frac{\lambda}{H}$$  $$G = 6 \frac{BH}{\lambda^2}$$

Fig. 1.19  Approximate formulae for an antenna with a parabolic reflector

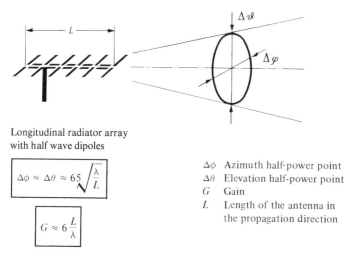

Longitudinal radiator array
with half wave dipoles

$$\Delta\phi \approx \Delta\theta \approx 65\sqrt{\frac{\lambda}{L}}$$

$$G \approx 6\frac{L}{\lambda}$$

$\Delta\phi$  Azimuth half-power point
$\Delta\theta$  Elevation half-power point
$G$  Gain
$L$  Length of the antenna in the propagation direction

Fig. 1.20  Approximate formulae for a longitudinal radiator array

The single radiators can be located longitudinally, relative to the direction of propagation, or transverse to it. In such instances they are termed 'endfire arrays' (Fig. 1.20) and 'broadside arrays' (Fig. 1.21). The individual radiators can be excited by a distributor network (active). For endfire arrays this can also be achieved by mutual coupling of active and passive radiators. The resultant polar diagram is dependent not only on the geometrical arrangement of the array, it is equally dependent on the amplitude and phase distribution of the power supplied to the individual radiators.

Other possible types of directional antenna, for instance lens antennae, are not discussed here since they have not so far been practically applied to secondary radar techniques.

The radiation properties of antennae are graphically represented in the form of directional diagrams for any specified plane. Thus, there exist both horizontal and vertical polar diagrams, such as have already been used to represent the omni-directional antenna in Fig. 1.15. The diagram is usually given in polar or rectangular coordinates (Fig. 1.18).

The desired result achieved by such a directional antenna is usually called the *main lobe*, and is defined by its gain relative to an isotropic emitter, as well as by its beam width (its width at the 3 dB point) in degrees. This is the angle subtended by the half-power points.

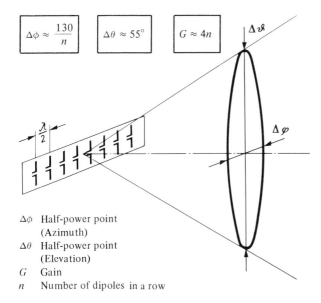

- $\Delta\phi$ Half-power point (Azimuth)
- $\Delta\theta$ Half-power point (Elevation)
- $G$ Gain
- $n$ Number of dipoles in a row

Fig. 1.21 Approximate formulae for a transverse radiator array (a line of dipoles with $n$ dipoles and a reflector)

Although it is not entirely avoidable, any radiation in other directions is undesirable. Such radiations are known as *side lobes* or *minor lobes*. A special instance of this is the *backward radiation* in the direction directly opposite to that of the main lobe. When describing an antenna diagram the largest of these spurious emissions is generally given as a ratio relative to the main lobe, in decibels, and known as side-lobe attenuation.

There exists a very close relationship between the directional effect and the mechanical dimensions of the antenna, or the number, $n$, of separate radiators. For rough calculations the approximations given in Figs 1.19–1.21 can be used. These formulae are only accurate to approximately $\pm 40\%$.

### 1.5.3. Matching and polarization

It has already been explained that the purpose of the antenna is to produce a complete transformation of the power supplied to it by the guided wave into a free-space wave and vice versa. During this process the loss of radiated power is not as critical as the fact that some portion of the power will be reflected at any mismatch. The reflected wave will thus be superimposed on the next forward wave, so that the information will be deteriorated, e.g. with pulse modulation this is the cause of *echo pulses*.

If power oscillators are used for the transmitter stage, as is usually the case in nearly all the responder units, a mismatched aerial can also have another effect. In this instance through the mismatch of the antenna to the feeder line, the resonator in the oscillator, which determines the frequency range, becomes more or less seriously mistuned (Fig. 1.22) which results in a frequency shift of the transmitter oscillator.

The voltage-standing-wave-ratio (VSWR) is generally used to describe the antenna matching. This factor is defined as the ratio of the maximum voltage $V_{max}$ to the minimum voltage $V_{min}$ on any line.

$$\text{VSWR} = V_{max}/V_{min} = s$$

The reciprocal of the voltage-standing-wave-ratio is known as the matching factor, $m$, and is defined as:

$$m = V_{min}/V_{max}$$

The modulus of the reflection factor $r$ is obtained from the quantities given above

$$r = \frac{s-1}{s+1} = \frac{1-m}{1+m}$$

In secondary radar installations a minimum requirement is specified for the impedance matching of the aerial for the reasons stated above. The maximum permissible value for the voltage-standing-wave-ratio is generally $s \leqslant 1.5$.

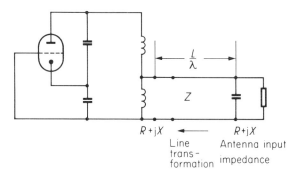

Fig. 1.22 Effect of an incorrectly matched load resistance on the natural frequency of the transmitter stage

A further property of antennae has not yet been discussed; namely, the polarization of the radiated field. In principle this polarization can be chosen in any desired direction provided that the transmitter and receiver positions are coordinated the same. Because an omni-directional radiation pattern in azimuth can more easily be realized for aircraft antennae it has been agreed that vertical polarization will be used in SSR system.

### 1.5.4. Coordinating the directional interrogator antenna with the primary radar antenna

The coordination of the azimuth angle between the primary and secondary radar installations can be achieved in many various ways. The choice usually depends on the type and structure of the primary radar system. In actual practice the following different solutions are generally used

- (i) A common pivot mount supports two separate antennae, one for the primary radar and one for the secondary radar interrogator (Fig. 1.23). In such instances, a linear *array antenna* e.g. an array of dipole radiators, is generally used for the secondary radar.
- (ii) A pivot mount supports a common reflector illuminated by two different primary feeds (Fig. 1.24).
- (iii) The two antennae are located on separate mountings, but their rotation is synchronized by a servo-control mechanism. This is termed the *slave antenna* (Fig. 1.25).
- (iv) The two antennae are set up separately, and rotate independently of one another. The coordination of the signals is achieved by using a

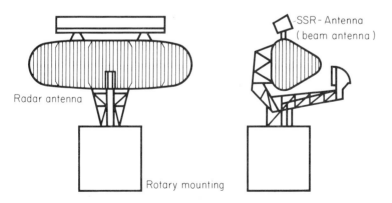

Fig. 1.23  Radar antenna supporting an SSR antenna

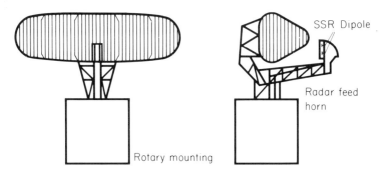

Fig. 1.24  Radar antenna with SSR emitters

display unit which has a deflection system allowing for the information from each antenna, to be displayed in turn.
(v) The two antennae are separately set up and rotate independently of one another. The information from the primary and secondary radars is processed in separate extractors, and is only correlated during the course of the electronic data-processing.

It is difficult to obtain good agreement between the primary and secondary radar radiation patterns. For economic and technical reasons it is sometimes necessary to make use of a common pivot mount as well as the reflector of the primary radar. This second requirement can only be satisfied if both systems operate within the same frequency band, since the size of the antenna and the directivity of a directional antenna are directly dependent on one another.

If, for design reasons, it is absolutely essential to base the dimensions of the secondary radar antenna upon the size of the primary radar antenna, then

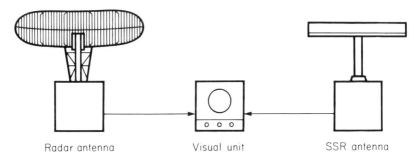

Fig. 1.25  Separate radar and SSR antennae

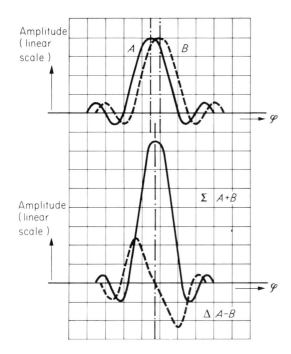

Fig. 1.26 Radiation pattern for a monopulse antenna arrangement

this means that the beaming of the secondary radar signal will be worse than that for an S-band radar by a factor of 3, and will be worse by a factor of 10 than that of an X-band radar.

In this respect an alternative is offered by the *monopulse technique* (Fig. 1.26). The following explanation will enable the essential details of this process to be understood – on the assumption that it is limited to a single plane (Fig. 1.27). The output signals from two directional antennae situated close together are connected, for example to a ring-hybrid loop, in such a way that the two channels provide the sum and the difference of the two antennae signals. The sum and difference signals are then connected to two separate receivers. The summation diagram gives the general direction of the energy received, with an amount of directivity which is roughly related to the geometrical dimensions of both antennae.

With the difference diagram, however, it is possible to determine a minimum amplitude in the direction of the axis of the loop between the two antennae and a phase reversal of 180°. This produces a very distinct criterion for the accurate determination of the target position. The outputs of the two receivers are again connected together in a monopulse evaluation circuit in

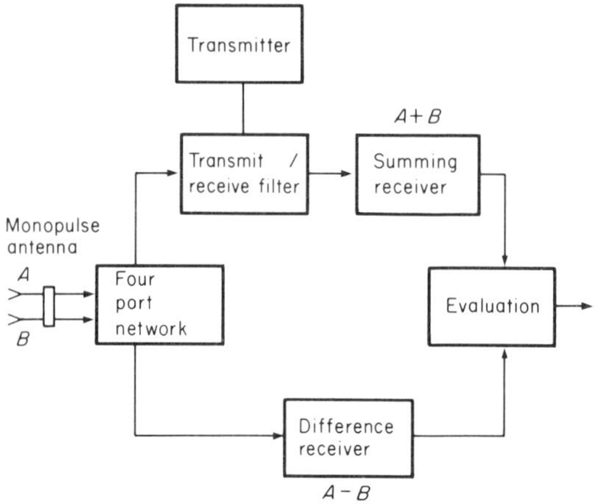

Fig. 1.27 Block diagram for a monopulse receiver

which the exact direction of the target is determined from the sum and difference diagrams. With the usual monopulse receiver, beam-sharpening factors of approximately 2–10 can be achieved.

A further advantage of this technique is the fact that the effective breadth of the display is no longer dependent on the receiver field strength, or the amplification characteristic of the receiver, but is only dependent on the ratio of the RF signals from the sum and difference output of the monopulse antenna. If an IF amplifier with a logarithmic transient response characteristic is used, its output will then correspondingly produce the difference between the sum and difference diagrams. This difference and, consequently, the breadth of the display, is the same at all ranges for a target with a constant response transmitter power. The breadth of this display can of course be affected by external control signals for special applications.

## 1.6. Side-Lobe Suppression (SLS)

During the description of directional antennae some reference was made to the phenomenon of side lobes. In all radar processes the result of these side lobes is that the determination of the angle becomes ambiguous when the reflected signals are strong.

In secondary radar processes this has a particularly disturbing effect since, in contrast to primary radar methods, the side-lobe attenuation only has

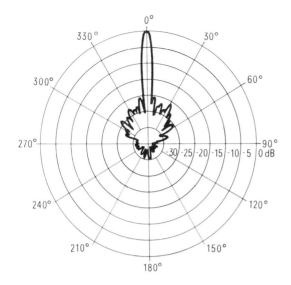

Fig. 1.28 Typical diagram of a directional antenna in polar coordinates

reference to one direction of propagation (Fig. 1.28). The result of this is that any aircraft in the nearby vicinity will be interrogated, not merely by the main lobe, but also by the side lobes. Such aircraft will no longer be displayed unambiguously on the screen. Countless other responses will also appear on the screen at incorrect azimuth angles. In an extreme instance the display may show as a complete circle, without any indication of direction. This is termed 'ring around'. There are several methods of side-lobe suppression (SLS methods) besides that of using larger and more expensive antennae with very small side lobes.

### 1.6.1. Interrogation path side-lobe suppression (ISLS)

The suppression of side lobes on the interrogation path depends upon the fact that, in addition to the actual interrogation which is emitted by a directional antenna, another signal from an antenna with a different radiation pattern provides a reference level (Fig. 1.29). The transponder then determines, by means of a level comparator, whether the signal is due to a main or to a side lobe, so that interrogations by side lobes get no response.

By means of the *two-pulse method* (which, in special instances, is still used in Great Britain), the first pulse, $P_1$, is fed to a directional antenna with

47

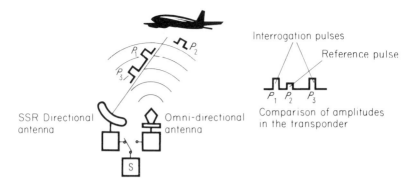

Fig. 1.29 Interrogation path side lobe suppression

lower directivity and a lesser gain, whilst the second interrogation pulse, $P_3$, is supplied to an antenna with higher directivity and a high gain. The criteria for the main lobe is, in this instance, the equality in the level of the two pulses (Fig. 1.30). Since the $P_1$ and $P_3$ pulses are to be of different power, if the two antennae with their greatly differing gains are to have equal field strengths in the main direction, they are generally produced in separate transmitters or transmitter stages.

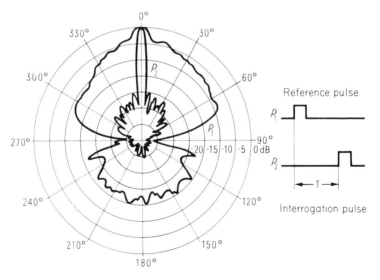

Fig. 1.30 Antenna diagram for double pulse method of side lobe suppression

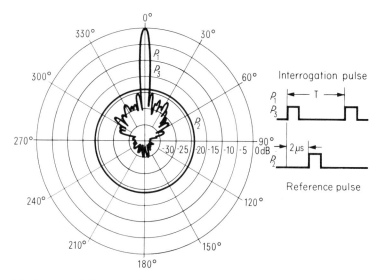

Fig. 1.31 Typical antenna diagram for the triple pulse method of side lobe suppression

In the *three-pulse method*, recommended as the principal method by ICAO, the two interrogator pulses $P_1$ and $P_3$ are emitted by the directional antenna, whilst a special control pulse, which follows the first interrogation pulse $P_1$ after an interval of 2 μs, is sent from an antenna with an approximately omni-directional characteristic (Fig. 1.31). The control pulse $P_2$ is supplied, together with the interrogator pulses, $P_1$ and $P_3$, from the same transmitter. By means of a fast RF power switch, the separate pulses from the group of pulses, $P_1$, $P_2$, $P_3$ at the output of the transmitter are switched to the appropriate antennae.

Interrogations by side lobes are then distinguished by the fact that they are received at a lower power level than that of the control pulse. If the equipment tolerances are taken into consideration, the relationship shown in Fig. 1.32 is obtained.

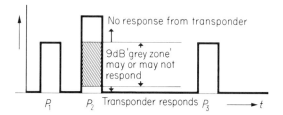

Fig. 1.32 Tolerances in levels for the $P_2$ control pulse during the triple pulse method of side lobe suppression

If $P_2 \geqslant P_1$, then the answer must be suppressed. If $P_1 \geqslant (P_2 + 9\text{ dB})$ the response must be produced. In the intermediate zone, which is 9 dB in breadth (the grey zone), an answer may or not not be produced.

A special variation of the ISLS method involves the use of a monopulse antenna when transmitting. The interrogation pulses $P_1$ and $P_3$ are emitted via the summation diagram, whilst the reference pulse is emitted via the difference diagram. By this means one can not only do without a special

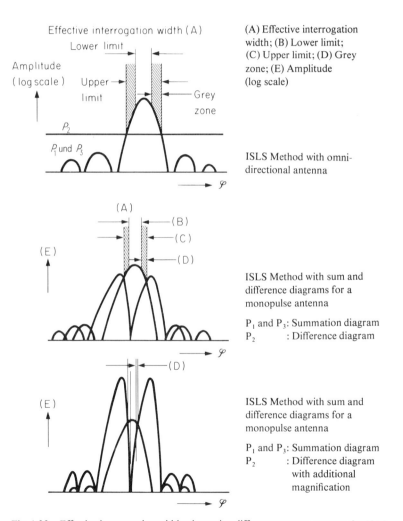

(A) Effective interrogation width; (B) Lower limit; (C) Upper limit; (D) Grey zone; (E) Amplitude (log scale)

ISLS Method with omni-directional antenna

ISLS Method with sum and difference diagrams for a monopulse antenna

$P_1$ and $P_3$: Summation diagram
$P_2$ : Difference diagram

ISLS Method with sum and difference diagrams for a monopulse antenna

$P_1$ and $P_3$: Summation diagram
$P_2$ : Difference diagram with additional magnification

Fig. 1.33  Effective interrogation width when using different antennae systems for ISLS

omni-directional emitter, one also obtains, almost automatically, the ideal agreement of the vertical diagram of the reference level antenna with that of the directional antenna.

If this method is extended, then, with the aid of the ISLS method, there is the possibility that already on the interrogation path the effective width of the main lobe can be reduced. The difference diagram stresses this with its higher transmitter power than the summation diagram. It is clear, from Fig. 1.33, that in this way the angular zone, in which $(P_1$ and $P_3) \geqslant P_2$ (and in which, therefore, a valid main-lobe interrogation occurs) is very much smaller.

### 1.6.2. Reply-path side-lobe suppression (RSLS)

In the process for suppressing side lobes on the reply path, all the responses are received by the ground station on two separate channels. In one channel they are received from a directional antenna, and in the other from an omni-directional antenna. The two receiver outputs are connected to an amplitude comparator with the result that only the signals received from the main lobe are displayed (Fig. 1.34).

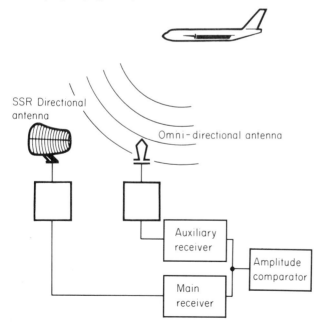

Fig. 1.34   Reply path side lobe suppression

In actual practice the two methods of side-lobe suppression, ISLS and RSLS, are used simultaneously.


**1.7. Fruit and Defruiters**

Section 1.1 has already shown how a secondary radar system operates when using the same channels. Multiple use of the same channel is facilitated by the fact that the pause which follows each short burst of activity is some hundreds of times longer than the burst. Because of the fact that, at the ground station, the interrogation, transmission and reception of the responses is performed with a directional aerial, there is a further substantial reduction in the area of space that is, at any time, being scanned. On the other hand, with the extended use of secondary radar systems, the number of ground stations and aircraft equipped with transponders is continually increasing, with the consequence that the probability of mutual interference is also increasing. This type of interference on a secondary radar system is called *fruit*. This implies interference during reception, caused by the fact that one interrogator station, A, will receive from an aircraft not only the responses to its own interrogations but also those responses which the aircraft is transmitting in reply to the interrogations

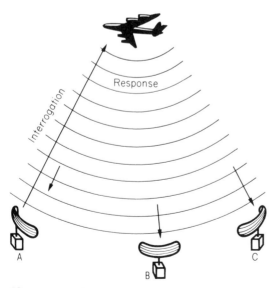

Fig. 1.35 Fruit, interference from other ground stations by responses emitted via one omni-directional radiation pattern

from other ground stations B, C, etc. (Fig. 1.35). Thus, it is not only the required answer which appears on the radar screen at station A. A profusion of replies having no connection with the actual interrogation from station A are scattered over the whole radar screen (Fig. 1.36). The ground station at A is thus faced with the problem of selecting the very small number of correct responses from the total. Quite a few years ago, in some of the focal points for air traffic, for instance London, a fruit density of 2000 replies per second was detected, whilst in the area of New York this value was 10 000 to 20 000 fruit per second.

A technical device known as a *defruiter* is one means of avoiding this problem. This device involves a synchronous filter which checks all the responses received to see if they are synchronized with their appropriate interrogation recurrence frequency, and suppresses those responses which are not in synchronism with their associated interrogation recurrence frequency, i.e. those responses triggered by the interrogations of other ground stations operating with a different interrogation recurrence frequency. Differences in recurrence frequency of the order of $10^{-3}$ are sufficient for this purpose. This test for synchronism can be so arranged that each response signal is stored on its arrival for exactly one interrogation period, is stretched to form an 'acceptance gate' for an allowance of the tolerances, and is compared with the response to the next interrogation

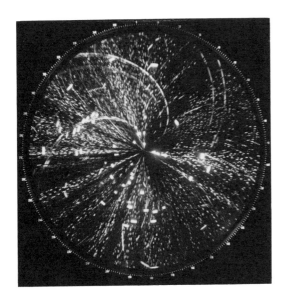

Fig. 1.36 Radar screen showing interference by fruit

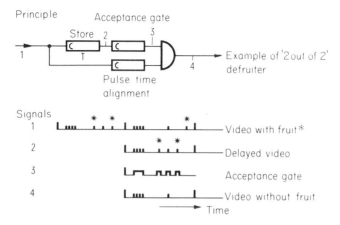

Fig. 1.37 Defruiter T Store for a period with a duration T

(Fig. 1.37).* This comparison of two responses, the incoming response with the stored response, represents the simplest evaluation criterion and is called a '2 out of 2' defruiter. It can readily be imagined, particularly if the fruit density is high, that an interference pulse may occur accidentally in the subsequent reception period within the same range increment. Such an interference pulse, which produces a synchronism for a single period obviously cannot be inhibited by a '2 out of 2' defruiter. However, where the fruit densities are fairly high it is possible to have twice, or thrice, this amount of storage thereby achieving the criteria known as '3 out of 3' or '4 out of 4' defruiters (Fig. 1.38). As will be shown later, these higher criteria provide a far better means of blanking the interference. In a defruiter, however, it is fundamentally impossible to avoid the loss of useful information. To reduce this it is possible, in any region of low or average fruit density, to use less strict criteria as for example '2 out of 3'.

The following *types of storage* have to be considered.

(a) *Delay lines*: When using delay lines it is assumed that the interrogation recurrence frequency can be derived from the delay line. Only under this condition will the storage time and interrogation period be accurately equal. They are not therefore very well adapted for universal use as defruiters.

---

* When discussing the methods of electronic data processing the terminology and symbols of German industrial standards[7] are generally used. A more detailed explanation would be beyond the scope of this book.

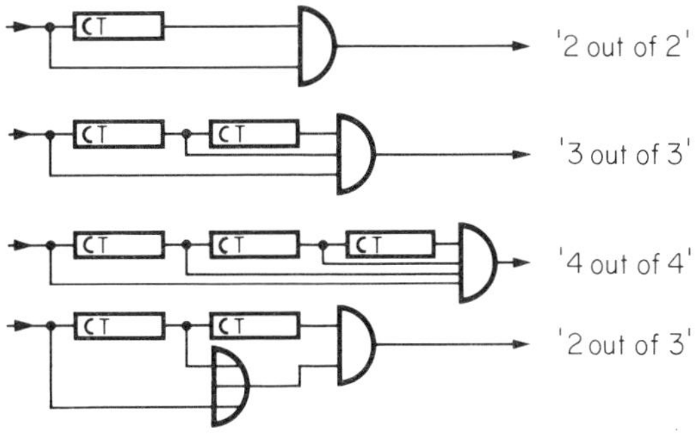

Fig. 1.38 Logical circuits for various defruiter criteria

'2 out of 2'

'3 out of 3'

'4 out of 4'

'2 out of 3'

(b) *Storage tubes*: In storage tubes the storage time is determined exclusively by the externally connected deflector function. They can therefore be easily adapted to any given conditions, i.e. they are universally applicable. With the older types of tubes, two storage tubes were required for each interrogation mode for the purposes of alternate 'write in' and 'read out'. A later type of tube makes possible simultaneous writing and reading, including the logical function of a '2 out of 2' defruiter. All storage tubes suffer from the inherent disadvantage that they are analogue devices and are, consequently, subject to certain limitations regarding accuracy, however costly the maintenance that they receive. Moreover, it is difficult to predict the working life of any single tube since this is determined by the signal-to-noise ratio produced by the storage layer.

(c) *Magnetic stores*: Magnetic stores belong to the group of digital storage devices. With such stores it is generally true that only two conditions, 0 or 1, are possible in each storage position. To store signals which occur in a temporal sequence, the time axis is divided into small steps denoted time increments or, in the radar process, directly called increments of range. A special storage position is then allotted to each of these steps.

If defruiters are digital then the required storage capacity becomes

$$K = (MR6.66/t)(k-1)$$

where $K$ is the required number of storage positions (bits); $M$ is the number of interrogation modes involved; $k$ is the evaluation criterion (e.g. $k=2$ for '2 out of 2'; $k=3$ for '3 out of 3'); $t$ is the range increment in microseconds and $R$ is in kilometres.

By carefully selecting the range increment it is possible to keep the quantizing error sufficiently small, with tolerable technical resources.

In magnetic stores the costs per storage position decrease as the storage capacity increases. Consequently, they are principally used in large defruiters, i.e. where there is a multi-mode interlace of interrogations, requirement of high resolution, high evaluation criteria and coverage of great ranges.

(d) *Shift registers with integrated circuits:* One of the results of the development of integrated circuits is, that, in the instance of simple defruiters, i.e. those operating over a small range, dealing with a single mode of interrogation and using a '2 out of 2' criterion, it is possible to produce remarkably economic units with shift registers which make use of integrated-circuit techniques. In this respect interesting developments can be expected in the sphere of metal-oxide semiconductor (MOS) integrated circuits.

Some further concepts have still to be explained to assess defruiters.

The width of the acceptance gate (Fig. 1.37) indicates the extent to which the individual pulses may deviate from their ideal position for 100% acceptance. These deviations are due, on the one hand, to the movement of the target between two interrogations, and also to unavoidable deviations in the timing of the interrogations and responses (jitter) which are fully acceptable according to the prescribed standards. Further, with digital defruiters, the digitizing error must also be accepted by the acceptance gate. The consequences of an unduly-narrow acceptance gate would be the suppression of correct answers.

The *rejection characteristic* indicates the probability of the rejection of an unsynchronized pulse: a pulse, therefore, which is outside the limits of the acceptance gate.

The *transient response* between acceptance and rejection, the *acceptance characteristic,* provides some indication of quality (Fig. 1.39). With defruiters using delay lines and storage tubes, this can be very easily realized in an ideal form. Some deterioration in the operating characteristic, however, has to be accepted since the practical use of digital defruiters is increasing due to their simpler maintenance and higher operational reliability.

When choosing the *evaluation criterion* it is most important that interrogation and response probability of the system can be assessed. This value is generally known as 'round reliability'. Figure 1.40 shows the effects of the various criteria in a practical situation where there is, under certain circumstances, an unusually high loss of information.

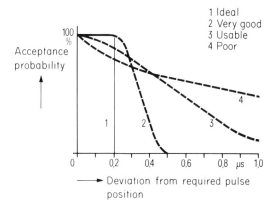

Fig. 1.39 Assessing the acceptance and rejection tolerance of defruiters

1 Ideal
2 Very good
3 Usable
4 Poor

The example assumes the existence of a ground station which scans at 10 rev. min$^{-1}$, and where the antenna main lobe has an effective width of 3°. At an interrogation frequency of 400 interrogations s$^{-1}$ 20 interrogations hit the target for each sweep of the antenna. If the round reliability is 100%, there should be 20 responses. If the round reliability is reduced, e.g. to 80% or 60% then only 16 or 12 responses will be received per sweep of the antenna, as illustrated in the first line of Fig. 1.40.

The subsequent lines clearly illustrate how, through the mechanism of the various defruiter criteria, useful information is lost with each gap in the information. The number of useful answers is indicated at the end of each series of interrogations. It thus becomes evident that with a '4 out of 4' criterion, and a reliability of 60%, only *one* response from 20 interrogations is displayed.

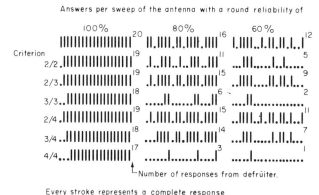

Fig. 1.40 Assessing the defruiter losses of defruiters

Obviously this loss must be taken into consideration if it becomes necessary to assess the possible extent to which interference can be suppressed, normally expressed by the defruiter efficiency, $\eta$,

$$\eta = 1 - \left(\frac{n_a}{n_e}\right)$$

where $n_a$, $n_e$ are respectively the number of fruit answers at the output and input. Figure 1.41 illustrates this for idealized conditions.

These observations make it clear that, although fruit can, of course, be suppressed by various technical devices, the extent of this suppression is very closely connected with the loss of useful information. It would be much wiser, therefore, to keep the fruit density as low as possible right from the start.

The following measures can contribute to this end:

(a) Establishing the least possible number of interrogator units. It is usually possible for a single interrogator unit to supply several users (air traffic control stations) with information.
(b) Low interrogation pulse-repetition frequencies.
(c) Target controlled interrogation such that the interrogations only occur within the angular sector in which a target has already been discovered by the primary radar.
(d) Perfecting every measure for the suppression of side lobes.

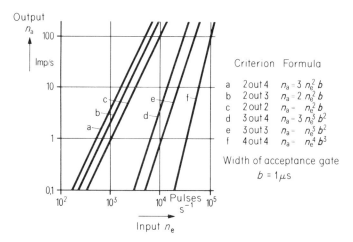

Fig. 1.41   Evaluating defruiters by the interference pulses at the output

## 1.8. Garbling

Every reply occupies a period of 20.3 μs, and an additional 4.35 μs may be required for the transmission of the SPI pulse. At the rate of propagation of waves in free space (for both the outward and return journey), this total period of 24.65 μs corresponds to a range of approximately 3.7 km. If two aircraft are simultaneously within the detection range of the interrogating antenna, and if the difference in their slant ranges to the target is less than 3.7 km, then the pulse telegrams from the two responses will, to some extent, coalesce in the ground-station receiver. This is a matter of some importance, since the beaming properties of secondary radar interrogation antennae are usually only of high quality in azimuth and not in elevation. Consequently, particularly when a large number of aircraft are staggered vertically, this may lead to confusion in the coding – to 'garbling' (Fig. 1.42).

In such instances the following distinction must be made. With *non-synchronous reply-code overlap* the overlapping of the responses is such that the two time scans do not coincide. Such responses, being within the resolving power of the decoder can then be separately decoded. With

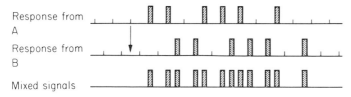

The time rasters fall into gaps and are separable

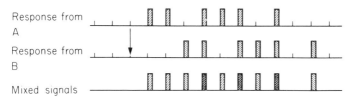

The time rasters partially coincide so that the responses produce a new false code which must be suppressed

Fig. 1.42  Garbling

*synchronous reply-code overlap* two, or even more, responses may overlap in such a way that they have a common time scan. It is thus no longer possible to determine, at the receiver output, whether an individual pulse belongs to one particular reply pulse train or to another. In such instances any decoding must be automatically suppressed.

## 1.9. Round Reliability of the System

Practical experience reveals that each interrogation does not always produce a response which gets through all the stages up to the actual display of the result. It frequently happens, for various reasons, that a few of the responses are missing. The following are usually the important causes of the failure.

### 1.9.1. The transmission channel

During the discussion on characteristics in Section 1.1, the somewhat simplified statement was made that, if the interrogator unit had a transmitter power of 1.5 kW and the transponder had a receiver sensitivity of $-75$ dBm, a range of 370 km was attainable. In actual fact, of course, there are a whole series of variable quantities which affect this result; for example, cable losses and the gains of the antennae both on the aircraft and on the ground.

The flow of complete information in any secondary radar system must be examined in both the interrogation and response paths. In the following example, only the interrogation path is taken into consideration. As the

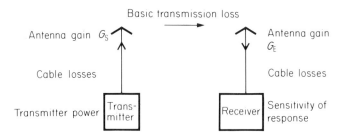

Fig. 1.43 Energy balance of a secondary radar system for the interrogation path

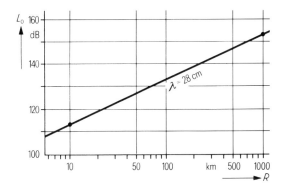

Fig. 1.44 Basic transmission loss for propagation in free space where $\lambda = 28$ cm

$L_0 = 93 + 20 \log R$

$L_0$ = Basic transmission loss in dB
$R$ = Range in km

conditions for the response path are very similar the same methods of calculation are nevertheless applicable.

The effects of the most important components in the system are represented diagrammatically in Fig. 1.43. In this instance the following equation can be applied:

$$P_{t_1} - P_{r_2} \geq L_0 + \alpha_1 + \alpha_2 - (G_1 + G_2)$$

The amount by which the transmitter power $P_{t_1}$ exceeds the response threshold of the receiver $P_{r_2}$ must be greater than the sum of all the losses over the transmission distance. The antennae gains, $G$, can be used in the calculation as a corresponding increase in the effective power radiated by the antenna.

The fundamental attenuation during transmission, $L_0$ (in decibels), when there is no interference in the propagation of waves in free space, depends on the wave length and the range $R$.

$$L_0 = 20 \log (4\pi R/\lambda)$$
$$= 20 \log (4\pi/\lambda) + 20 \log R$$

For the frequencies in a secondary radar system this relationship can be represented, graphically, as a non-dimensional equation, as illustrated in Fig. 1.44. There are additional losses – chiefly due to humidity – but, in any approximate calculation, this additional attenuation can be disregarded at the frequencies used in secondary radar systems.

In the typical secondary radar installation used for air traffic control the following values are obtained for the interrogation path.

| | | |
|---|---|---|
| Transmitting power | $P_{t_1} = 1500$ W | 61.5 dBm |
| Receiver sensitivity (transponder) | | |
| Level of transponder response | $P_{r_2}$ | −75 dBm |
| Gain of interrogation antenna | $G_1$ | +21 dB |
| Gain of aircraft antenna | $G_2$ | 0 dB |
| Effective excess of transmitter power above transponder response level | | 157.5 dB |
| Cable losses at the ground station $\alpha_1$ | | 3 dB |
| Cable losses in the aircraft installation $\alpha_2$ | | 3 dB |
| Power left to overcome basic transmission loss | | 151.5 dB |

This attenuation corresponds to a range $R$ of approximately 800 km

The values for the response path give a similar balance.

This statement does not, however, include any reserves for deviations in the characteristics of the system. Conversely, it may be said that, to cover a range of 370 km (nominal range), a reserve of 8.5 dB is available. It is however, questionable whether this reserve is sufficient.

It is generally impossible to give absolute figures for these characteristics. All quantities, in engineering, are subject to permissible tolerances. Thus the nominal value for the transmitter power is 1.5 kW. In actual fact with new tubes it is somewhat higher but, after a considerable period in operation, any reduction in transmitter power is only detected by the monitoring device if the nominal transmitter power is reduced by more than 1 dB. Similar remarks hold good for the receiver sensitivity. In the instance of antennae in particular, large fluctuations in the assumed values for the gain must be taken into consideration. The reason for this is partly due to the fact that, generally, the data are only known for a single preferred plane and partly because of the formation of lobes in the vertical diagram, as happens in the instance of the interrogating antenna through reflections at the earth's surface, and in the instance of the responder unit antenna through the effect of the aircraft frame (fuselage etc.).

When covering aircraft at low altitudes over a good conducting surface (the sea) this effect can be very aggravating.

When considering the action of the ground antenna, the method of scanning has to be taken into consideration. The assumption of maximum gain in the power balance is only justified if the antenna, whilst dwelling on the target, is at its radiation maximum, as is the instance with target-tracking units. In surveillance radar installations, however, the antenna lobe only scans the target for a very short period. The time that the antenna dwells on the target, and, consequently, the number of interrogations (hits per scan) also depends on the extent to which an interrogation can be made below the maximum gain. A more accurate relationship is given in Section 1.9.4, but the nomogramm in Fig. 1.47 makes it simple to gauge the amount that must be deducted from the maximum gain of the antenna to produce the power balance.

In the instance of the aircraft antenna it is necessary to take account of the fact that the viewing angle (aspect) at which an aircraft is covered from the ground station, is also subject to a definite probability distribution. Thus, although the aircraft antenna has a very low gain (theoretically a null) in its perpendicular axis, the moment during which it flies directly over the ground station will only be a negligibly small fraction of the total period of coverage along the flight path. (Note that when producing the probability distribution for the antenna gain, this can be taken into consideration provided that different weighting is allotted to the different aspect angles.)

Of any single quantity, therefore, we only know within what values it will *probably* lie, and the value it will assume for any specified probability. It is therefore clearly advisable to make use of statistical methods.

An attempt will therefore be made to give a brief insight into the fundamental ideas and the basis of such a statistical method of treatment.

The relationship for the power balance on the interrogation path is shown in Fig. 1.45. Each quantity, which has any influence, is represented there by a function which indicates the probability, $p$, of a specific value, e.g. of transmitter power, being available. This graphical representation is termed the probability density. By a mathematical operation known as 'convolution' it is possible to obtain a new probability density from several separate probability densities.

A calculation has been made for an example with a variable parameter (transmitter power $P_{t_1}$), and this has also been represented graphically. This graphical representation of the probability distribution (Fig. 1.46), in contrast with the previous diagrams of probability density, uses a new set of coordinates, a so-called probability grid. This states the probability, $p_1$, with

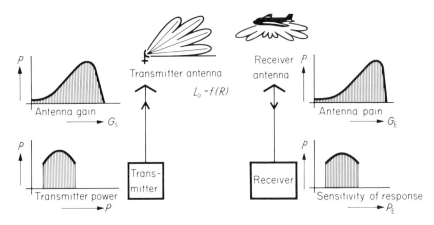

Fig. 1.45 Statistical distribution of the system values in a secondary radar system (interrogation path)

which, in a known system with a transmitter power $P_{t_1}$, a given range, $R$, will be attained or exceeded. Thus it would be pointless to provide a 99.9% probability for a single contact since this, besides being uneconomic, would where the propagation conditions were especially favourable, also result in an excessive range (with interference, as a consequence, in areas very far away). It is, on the other hand, sufficient that, for several efforts made in sequence, there should be a steep rise in the probability that one of these efforts will be successful.

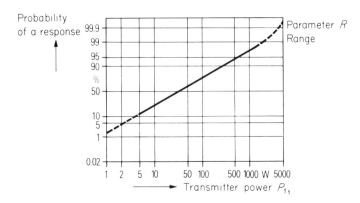

Fig. 1.46 Probability of a response from a transponder at a range, $R$, as a function of the transmitter power, $P_{t_1}$, of an interrogator unit (example)

If $p_A$ is the probability that an interrogation will successfully trigger the transponder, then the remainder is $(1-p_A)$. If a fresh effort is made, which is statistically independent of the preceding effort, then the probability that the transponder would be triggered at least once after the second attempt becomes

$$p_2 = p_A + (1-p_A)p_A$$

In a general form the probability $P_H$ that the transponder will respond at least once in $H$ attempts is

$$p_H = 1-(1-p_A)H$$

Here one must note a very substantial difference as compared with primary radar techniques.

In primary radar techniques the detection of an echo is made all the more uncertain because the signal-to-noise ratio is so low. However, this may be changed with the very next echo, i.e. the next 'hit'. Because of the noise even two consequent echoes are statistically independent of one another. Consequently, the probability of detection increases with the number of hits per scan.

The secondary radar process is operated with a high signal-to-noise ratio. Conditions scarcely change during a sweep of the antenna so that they are statistically dependent on one another. Consequently, an increase in the probability of a response can only be achieved by an integration over a period of time within which the propagation conditions have changed, due to the change in the position of the target; that is, in many cases only after a few revolutions of the antenna.

The data quoted are usually for the 90% probability of a response per single scan.

### 1.9.2. The transponder

The receiving of a valid interrogation does not always result in the triggering of a response. The absence of the response may be due to the following characteristics of the system:

(a) The transponder will only accept a fresh interrogation if the previous interrogation has been answered (decoupling of transmitter from receiver). In addition, after each response a dead time of 35–100 μs is introduced so that each participant in the secondary radar system has a chance to make connection with the ground station.

(b) One of the effects of the side-lobe suppression circuit (ISLS) is that the response to interrogations is also suppressed for a given period. This dead time, the ISLS dead time, usually amounts to $35 \pm 10$ µs.

(c) On board an aircraft, several radio systems are frequently operating simultaneously within the same frequency band. Frequently, for example, only a single aerial is provided both for the SSR transponder and for the DME range measuring system. To prevent mutual interference in these systems they are locked from one another by special *suppression signals*.

(d) The transponder has an *echo suppression circuit* to prevent distortion of the directly-received responses by the superimposition of responses received by reflection. The suppression of these echoes is achieved by reducing the receiver sensitivity for a brief period. Should the echo suppression circuit be triggered by an interference pulse then the true interrogation immediately subsequent to it would reach a transponder, the sensitivity of which has been reduced so that the probability of triggering a response would be correspondingly lower.

(e) The interrogation density which must be used for any reasonably realistic number of interrogator stations can be derived from statistical data. This frequency can then be used as the basis for dimensioning the mean power for the transponder transmitter stage. To prevent the transmitter stage being overheated by an unintentionally high number of interrogations an *automatic overload control* (AOC) is usually provided. There are two types of AOC:

  (i) When there is an overload, the receiver sensitivity is so far reduced that weak interrogator stations are no longer received. In this way the number of interrogations is reduced to such an extent that the permissible loading of the transmitter can no longer be exceeded.

  (ii) Although all the interrogations are received, only a certain percentage are permitted to trigger responses under overload conditions. This method again prevents the average number of responses exceeding a specified maximum limit.

  As regards the probability of a response from the system, both methods indicate that there will not be a response to every interrogation when the number of interrogations is excessive.

(f) There are occasions when the pair of interrogation pulses are so severely limited by the unduly low dynamic characteristics of the receiver that, at the receiver output, the interrogator pulses ($P_1$ and

$P_3$), and the SLS pulse, $P_2$, are of approximately the same size, so that the transponder reaction is to suppress any response just as though it were reacting to a side-lobe interrogation. This phenomenon is known as 'main beam killing'.

(g) If the dynamic characteristics of the receiver are too low, the message transmitted may be falsified (pulse distortion), should the signal level be very high. This will not only cause faults in the coding; it will also cause additional dead times – if these are triggered by interference pulses (squitter).

(h) Failures due to lobe switching. Ideally, the transponder antenna should have an omni-directional characteristic to be able to receive interrogations from all directions, and provide the responses. In practice, the actual fuselage of the aircraft has a powerful screening effect such that the antenna generally only covers a hemispherical space beneath the belly of the aircraft. With civil aircraft, which generally maintain a horizontal attitude this is not of very great importance. However, with military aircraft, special precautions have to be taken to ensure that identification can be made, even when the aircraft adopts unusual flying altitudes. A simple change to the upper hemisphere cannot be achieved by means of two antennae in parallel owing to the occurrence of very powerful mutual interference phenomena.

Some partial assistance can be obtained by switching between the two antennae (lobe switching), at a rate of a few Hertz, and thus covering every direction; but only, of course, with an approximately 50% probability of response.

### 1.9.3. The defruiter

The purpose of the defruiter, as described in Section 1.7, is to test whether responses from the same target range are repeated with the same timing as the interrogations. For this purpose every response is stored for a single interrogation period, and is then compared with the next interrogation. The number of responses filtered out in this way is of necessity always less than the original number.

### 1.9.4. Decoding the response

Failures also occur through excessive tolerances in pulse timing. These can be thought of as equipment errors, and will not be discussed here. In

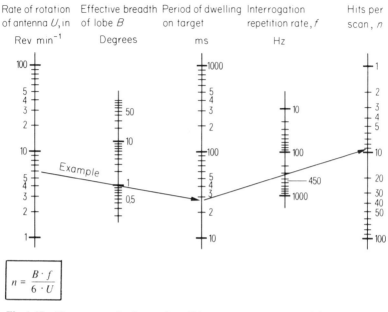

Fig. 1.47 Nomogramm for the number of hits on a target per sweep of the antenna as a function of the rate of rotation, the beaming and the interrogation frequency

contrast, a real systematic loss of responses is due to their suppression caused by a synchronous overlapping of the responses (garbling, see Section 1.8).

In conclusion, it can be stated that, as the density of secondary radar systems increases, the probability of a response from the system decreases. In a secondary radar system, however, the reliability of the data is ensured, not by any safety precuations within each reply but by the redundant multiple transmission of the same information. If this principle is to be effective, one essential requirement is that during each rotation of the antenna a specified minimum number of responses must be capable of being received in each interrogation mode from each aircraft.

When designing any installation a theoretical number of hits per scan (without losses) must be specified. This is dependent on the beaming of the antenna, the interrogation pulse repetition frequency and the rate of rotation of the antenna. It can be defined by the following equation

$$n = (Bf)/(6U)$$

where $n$ is the number of hits per scan (without losses); $B$ is the effective beamwidth of the antenna lobe in degrees (this is not necessarily equal to the half-power beamwidth $\Delta\phi$); $f$ is the interrogation pulse repetition frequency in interrogations s$^{-1}$; $U$ is the rate of rotation of the antenna, in rev. min$^{-1}$.

This relationship is shown in Fig. 1.47. An auxiliary scale makes it possible also to read the length of time that the antenna lobe dwells on the target.

To succeed in evaluating the information correctly (when the probability of a response is taken into consideration), one usually operates at 6–10 hits per interrogation mode.

## 1.10. Possible Development in Future SSR Systems

During these past years the SSR system has proved to be an extraordinarily valuable subsidiary aid in air traffic control and every effort has been made to cover all the users of any given air space with as few gaps as possible.

With the increasing number of partners, however, or, rather, with the increasing number of ground and responder units, the limits of the present system, which involves those problems, already indicated, of fruit, interrogation overloads, and garbling, are daily becoming more and more distinct.

Generally, all these interference phenomena can be traced back to the same cause; namely, to the method of scanning used by the SSR surveillance units that have so far been installed. As explained in Section 1.1, the interrogation so produced is a call 'to all aircraft'. The selection of the required response can only be made at the ground station, after a host of responses have been unnecessarily triggered. The resulting systematic interference phenomena were taken for granted and were subsequently eliminated by more and more complicated technical devices. Electronic data-processing (EDP) equipments which are used today to solve a vast range of complicated organizational problems may also be a means of handling these problems.

The basic features and potentialities of automatic decoding were discussed in Section 1.3.4. Since data-processing equipments can store, for subsequent processing, the complete air situation over a much longer period of time, the method of conducting the exchange of information with the ground station, and the design of the responder unit can be completely redesigned.

Another example of the amalgamation of EDP equipments with RF transmission equipment is the 'phased array', an electronically-controlled directional antenna. Such an antenna is an almost inertialess means of covering every point in space. Thus, completely new methods of scanning become available since it is now possible to jump from target to target and only to dwell on each target as long as is required to obtain a reliable exchange of information.

By taking advantage of these new technical possibilities the present SSR system could be so extended that it would be able to handle the enormous increase in air traffic to be expected by 2000 A.D., and to surmount all the problems of air traffic control that this will involve.

### 1.10.1. An address-coded SSR system – selective interrogation

An experimental programme known as ADSEL (*A*ddress *Sel*ective SSR System[8]) is currently in progress at the Royal Radar Establishment (RRE), Malvern, Worcestershire, England.

Under the control of the FAA (Federal Aviation Authority) various research institutes and firms in the U.S.A. are progressing with a proposed system known as DABS (Discrete Address Beacon System).[9]

Although the proposed systems for ADSEL and DABS were quite independently developed, they do have certain basically common features, the most important of which is the use of an address code; the use, that is of a selective interrogation, to which fact the names of both these programmes, ADSEL and DABS refer. It is almost certain that agreement will be reached on some common concept for the submission of a system to be standardized by ICAO.

The following pages will, consequently, only give a rough outline of the possible characteristics of a future SSR system without giving any particular details of the ADSEL and DABS systems which are both still in the development stage.

A significant confirmation of the value of the system is the tendency, during development, to construct the future SSR selective address system on an evolutionary principle; that is, to make it compatible with the existing SSR system with the result that, besides being able to continue to benefit from the outlays invested in the present system, it will also be possible to adapt the SSR system to suit air traffic densities which will vary regionally.

If the present frequencies of 1030 MHz and 1090 MHz are to be used for the RF transmission of interrogations and responses, then efforts will need to be made to ensure that the data format for the old and new systems are mutually compatible.

Thus, the following combinations must be considered for the interrogation direction.

(a) An 'old type' transponder is interrogated by an SSR mode and responds with an SSR code.

(b) An 'old type' transponder receives an address-coded interrogation to which it must not respond. This can be achieved in various ways. The interrogation can be modulated in such a way that the receiver cannot 'see' it. In the simplest instance, different methods of modulation could be used, such as phase- or frequency-shift keying. With such methods, however, it is necessary to be certain that unwanted responses will not be produced by the great number of transponders in the area. Therefore an address-coded interrogation, will start with a preamble containing the SLS criterion of the SSR system i.e. a pair of pulses with a 2 µs interval (Fig. 1.48). Any subsequent address code would then be rejected.

(c) A 'later type' of transponder receives a coded interrogation containing its own address. The address portion must be recognized, e.g. the synchronized group fed to a special decoder, and answered by a special reply.

(d) Another 'later type' of transponder receives an interrogation in one of the standard SSR modes. A normal SSR reply will be given in response.

The type of antenna to use with an address-coded SSR system is a problem which depends, finally, on the density of aircraft using the coverage area.

The electronically-controlled antenna undoubtedly offers the greatest freedom of choice as regards methods of scanning and the production of interference-free connections. The 'E-scan-antenna' (Fig. 1.49) (electroni-

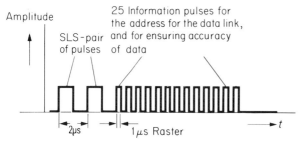

Fig. 1.48 Example of an address code in an SSR system with selective interrogation

cally controlled antenna) (Hazeltine Corporation, Greenlawn, New York, U.S.A.) developed as part of the DABS programme is not only capable of performing an electronically-controlled sweep in azimuth, it also has carefully dimensioned vertical radiation patterns for both directional and omni-directional radiation, the lower limit of which is made to 'follow' the contour of the horizon. This very considerably reduces ground reflections and, consequently, almost entirely suppresses the formation of lobes in the vertical diagram. During reception the antenna has a sum-and-difference pattern to permit the connection of a monopulse receiver.

In address-coded secondary radar systems the monopulse technique is of particular importance. With the selected method of scanning, the antenna lobe is directed to the azimuth position, of the aircraft under call, as predicted by the data logger. In actual fact, of course, the true location of the aircraft will not correspond exactly with the predicted value. In the conventional method of scanning, the antenna lobe traverses the target and, after a considerable number of hits per scan, subsequently determines the centre of the target. In the address-coded secondary radar system, however, an effort is made to work with a small number of responses; in the ideal example, with a single dialogue, i.e. one response to one interrogation. In principle, this can be achieved with a monopulse system. But, of course, noise and interference signals make it essential even in a monopulse system

Fig. 1.49 Combination of a conventional radar antenna with an SSR antenna with electronic beam scanning (electronically scanned antenna, E scan antenna)

that there should generally be a certain averaging, i.e. that there should be several responses to several interrogations.

Although the 'E-scan-antenna' illustrated is considerably more efficient and powerful, it is also much more expensive than the conventional directional antennae. Preparations are being made for scanning areas less densely covered with aircraft where use is normally made of conventional antenna rotating according to a rigid plan. In this instance the interrogations will be controlled by an EDP in such a way that the aircraft which happen at any moment to be within the antenna lobe will be selectively interrogated. If there are only a few aircraft within the breadth of the lobe this idea is perfectly practicable. Greater reliability can again be attained by interconnecting the information obtained from several such interrogator units over a greater area.

A combination of mechanical and electronic methods of traversing the beam provides a compromise between efficiency and costs. For this purpose a 'phased-array' antenna is used with a field of traverse limited to $\pm 15°$. This antenna is installed on a mounting which is mechanically rotated at a uniform rate, just as the antennae in the usual SSR installations. With this arrangement it is quite simple to make the lobe dwell for a longer period on sectors with a fairly large number of targets, or ignore those sectors without any targets, merely by superimposing a counter-rotating or a following electrical traverse on the uniform mechanically-produced rotary motion.

Now that the essential components of an address-coded secondary radar system have been described, one can imagine that it may operate in the following way.

Within a given acquisition period, the air space is scanned with a call to *all* aircraft in the usual way, and the information obtained is stored in a data extractor. In this phase it is also possible to deal with such aircraft which do not have transponders equipped for address-coded interrogations. As their number increases, the quality of such information will of course be reduced. The information, on the aircraft already covered by the store, now only needs to be up-dated at specified intervals.

Since the probable location of a target might be obvious from its previous position, the antenna can be aimed directly at the target, and a selective interrogation can be transmitted. The time that the antenna dwells on the target is selected to ensure that, under all transmission conditions, exactly the right number of dialogues occur to pass on the required information. The time required for this phase of 'aimed' interrogations is thus so small that it can easily be interlaced with the 'acquisition phase'.

### 1.10.2. The data-link capability

Because a uniquely-defined transmission route can be arranged from a ground station to a transponder, an address-coded secondary radar system offers even further possibilities. For a long time now, air traffic control organizations have been seeking a less expensive substitute for transmitting certain routine information through the radio-telephone channel. This system makes it possible to add coded air traffic control information to the actual information on identification. The SSR system is thus extended to become a *data link* (with a limited transmitting capacity).

### 1.10.3. The Synchro-DABS method

A special method of organizing the address-coded SSR system known as the *Synchro-DABS* method has been proposed. This method enables aircraft to give one another a proximity warning indication (PWI) by using the secondary radar system. To understand the Synchro-DABS system it is necessary here to give a more detailed explanation of the method used for measuring range with the help of uni-directional propagation.

In the radar systems so far described, the measurement made has always been that of the elapsed time between the emission of the transmitter pulse (or interrogation) and the return of an echo (or response). The range can then be calculated from the known rate of propagation of electromagnetic waves.

For a long time the only practical solution to this problem was an arrangement by which the beginning and the end of the time interval required to be measured was obtained at the same location. In fact, range can of course be determined from only one propagation path if the beginning and the end of this can be determined by data in absolute time. Until recently the practical application of this method for proximity warning indication purposes has been impossible for the simple reason that clocks of sufficient accuracy – generally atomic time standards – are far too expensive for inclusion in smaller aircraft. However, much simpler clocks, for instance quartz-crystal clocks, can also provide sufficient accuracy provided that they are regularly synchronized from the ground. This problem of synchronization has been solved by the Synchro-DABS system. In this system the interrogations are not emitted in a sequence with fixed timing. Each addressed interrogation is released at such a time that it will 'hit' the target at a previously agreed time (for example, at any specified second), and synchronizes the target clock, which must only act as an intermediate standard.

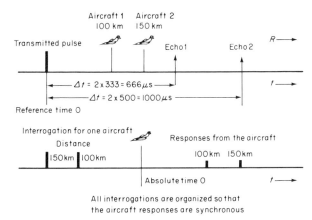

Fig. 1.50  Comparison of time sequence for a conventional radar installation (above) and for inquiry and reply with the Synchro-DABS system (below)

The replies from all the transponders sharing in the Synchro-DABS system are also produced in this way; i.e. they are always transmitted at very accurately-specified moments in time (see Fig. 1.50).

If it has an additional PWI unit, each of the other aircraft can receive responses from the rest of the transponders, determine their range and evaluate the identification and altitude signals transmitted. Data on range, change and flying altitude are combined with their own flight data. If there should be a possible danger of collision, the pilot is warned by being given the 'address' of the aircraft with which he is in danger of colliding.

### 1.10.4. Airport surface surveillance

Airport surface surveillance is another very attractive suggestion for the use of an address-coded secondary radar system, since it provides a further use for the transponders now available. The required accuracy in the location of a target cannot of course be obtained with the usual methods of scanning, the reasons for this being as follows:

(a) On aerodromes, a multiplicity of reflections and, consequently, of indirect propagations, have to be taken into account.
(b) For any tactical supervision system the rate at which data is up-dated must be very flexible.
(c) It would appear to be doubtful whether an aerodrome could provide a suitable position for the erection of a conventional antenna with an adequate power of resolution in azimuth.

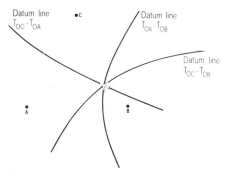

Fig. 1.51 Hyperbolic location system. To be measured: the delay-time difference of a reply between the transmitter 0 to be located and the receiving stations A, B, and C. The geometrical location, and therefore the datum line, for each pair of stations is a hyperbola. The intersection of the three datum lines shows the actual position of the transmitter

(d) Much greater accuracy in the location of targets is required than when monitoring the air space. The accuracy required should be better than the dimensions of the aircraft under surveillance.

It appears to be possible to construct a hyperbolic location system with a so-called 'whispering interrogation', an interrogator unit with a low-power transmitter, and several receivers located beyond the coverage area. In this instance, a hyperbolic location system means an arrangement whereby the same response is received from two distant receiver stations, and the difference in time in the receipt of the respective messages is measured. The locus for all transmitters which are received after the same difference in time is a hyperbola, the foci of which are the receiver locations. If a third receiver station is now added to the two already mentioned, then two pairs of receiver stations can be formed which can measure two sets of differences in the time of reception and so produce two hyperbolae as the loci of the transmitter. The point of intersection of these hyperbolae (datum lines) is the location of the transmitter (Fig. 1.51).

Since the basic principles of secondary radar techniques have now been discussed, the remainder of the text will give a more detailed description of the equipment used. As this description will be based on the engineering techniques in use today, many of the units described may, within the foreseeable future, become outdated. However, the majority of the applications are not dependent on the latest technical discoveries. Some quite old units are still satisfactorily fulfilling their purpose. On the credit side, they can frequently offer a wealth of operational experience, high reliability, and sound methods of operation. Consequently, some of these older units will be introduced when the occasion arises.

## 2. The Design of Interrogator Units

The chapter on *Basic Principles* which has already described the main units in an interrogator, has discussed the problems which this unit involves as regards the system, and has explained the fundamentals upon which the design of the interrogator depends.

Because the secondary radar unit is always operated in very close cooperation with the primary radar unit, both units are shown in a combined installation in a block diagram (Fig. 2.1) to show the common principles, the differences and the interfaces. Since it is very easy to define the functions and design of a simple primary radar unit (indeed, these can

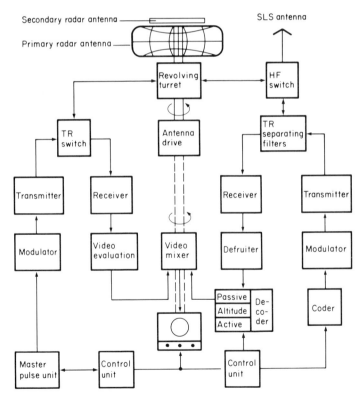

Fig. 2.1  Combined primary and secondary radar installation

be assumed to be generally understood), only the essential components will once again be briefly discussed.

To make the explanation easier to understand, numerical values are also quoted for the important units. These are the typical values for an S-band air surveillance radar, with a range of 200 km.

The *clock pulse generator* of a radar unit is the starting point for all the time measurements effected in the unit, and for the range measurements derived therefrom. This is the unit which determines the time at which the transmitter must emit its pulses, the period for which the sensitivity time control of the receiver (STC), which is dependent on range, must be adjusted, and the period for which the analogue measuring unit in the visual display must be triggered. In addition, there are auxiliary signals such as range markers. To understand the significance of the clock pulse generator it is also necessary to realize that the amplitudes and shapes of the triggering pulses generally have to satisfy very varied requirements, and that any delays caused by time shifts in any individual units must be compensated for in the separate units of the installation. The clock pulse generator coordinates all the timings relative to a temporal zero point.

The purpose of the *modulator* is to convert the trigger signal from the clock pulse generator into a signal suitable for controlling the radar transmitter. With the usual magnetron-controlled transmitter this process of modulation is achieved, principally, by switching the operating voltage on and off. The peak pulse power required of the modulated pulses amounts to several megawatts, as can be estimated from the transmitter power (efficiency!).

At a pulse repetition frequency of 400 Hz the pulse length is 1 µs. This corresponds to a duty cycle of 0.4%, so that the mean power is some kilowatts. The effects are simpler if radar units generally operate with constant pulse-repetition rates. It then becomes possible to make use of resonance effects and, further, there is sufficient time during the spaces for recovery processes, such as the recharging of storage capacitors or delay line networks.

The *transmitter* of a primary radar unit generally has a magnetron* as its RF generator. The stability of its frequency is not very high. This, however, is not very important since the receiver which recovers the echo signals at the same frequency, is finely tuned to the transmitter frequency by the

---

* The term magnetron defines a velocity-modulated valve which converts the modulated d.c. power supplied by the modulator into RF power. The frequency of the RF oscillation is thus determined by the mechanical dimensions of the magnetron.

COHO (coherent oscillator). At frequencies in the S-band (i.e. with wavelengths of about 10 cm) the transmitter has a peak pulse power of 1 MW.

The *TR switch* decouples the transmitter and receiver, if, as is usual, a common antenna is used to emit the transmitter pulse and to receive the echo signal. Because the transmitter and receiver frequency are essentially the same, they can be separated only by a switch. Advantage is taken of the fact that there is a difference in time between the transmitted pulse and the reception period. The time interval required to switch over from transmitting to receiving explains why more than one range is quoted for radar equipment. Besides the maximum range, there is also a minimum range. The essential properties of the switch are low insertion loss and high cut-off attenuation, i.e. low attenuation when 'on' and high attenuation when 'off'.

To prevent damage, for example, to the extraordinarily sensitive mixer crystal by cross-talk interference from the transmitter pulse (1 MW) at the input to the receiver, the switch must have a cut-off attenuation of at least 70 dB.

The *rotating joint,* which has several separate channels, connects the feed line from the stationary part of the installation to the rotating part of the antenna. The difficulty implicit in the radar channel is the provision of a low-loss transmission of high RF power with low and constant VSWR. The other channels for the transmission of secondary radar signals are easier to design.

The *antenna* of a radar unit comprises a feeder horn and reflector. A so-called fan-shaped beam is produced by the appropriate dimensioning of these items. In the horizontal plane of the directional pattern of the antenna the beaming is thus very good, corresponding to an angle of 2°, whilst the beaming in the vertical direction is such that all targets between the horizon and an angle of elevation of 35° can be covered, i.e. between the horizon and the *maximum permissible* flying altitude.

The purpose of the *antenna drive* is to orientate the end-on directional antenna, as required by the scanning programme; in this instance, therefore, to maintain its rotation about its vertical axis. Within certain limits the selection of the direction of rotation and the rotational speed can be remotely controlled by the radar operator. Being of robust mechanical construction, and having sufficient power in the drive, the regular operation of the drive is ensured even in stormy and icy conditions.

The device for electrically transmitting to the display unit the instantaneous azimuth position of the antenna (the 'electrical shaft') is an important component of the antenna drive.

The *radar receiver* accepts the echo signal reflected from the target via the antenna, the rotating joint and the TR switch. It is designed as a heterodyne receiver, in which the oscillator is automatically controlled by the transmitter frequency. The receiver sensitivity, in our example, is approximately $-110$ dBm. The intermediate frequency is usually about 30 or 60 MHz, and the bandwidth amounts to a few megahertz.

Very closely connected with the receiver are the various units for *signal processing*. Amongst these there is the so-called 'moving target indicator' (MTI). This recognizes the signals from a moving target from Doppler effect, which involves a frequency shift between the transmitter and echo frequencies. Echo signals caused by stationary targets causing interference in the air space being monitored can thus, to some extent, be suppressed without influencing the simultaneous display of useful targets.

Sensitivity time control (STC) is another way of controlling the received signal. Its purpose is to compensate for the received power which decreases so rapidly with increasing range $R$ (namely, by $1/R^4$). For this purpose the sensitivity on the transmission of the transmitter pulse – controlled by the clock pulse generator – is very much reduced, but thereafter increases as a function of time and, consequently, of the range, in accordance with the range law. The maximum sensitivity is again reached just before the maximum range of the installation. The radar video signal, which is made available, is the output signal from the receiver and signal processor.

The purpose of a *video mixer* is to combine the actual radar video signal, which represents the echo signal, with other information which also needs to be displayed; such as range markers or video maps, to simplify the determination of range azimuth and the orientation of targets on the radar screen. As regards this combined unit it is extremely important to have an input through which the passively-decoded SSR response can be allotted to the appropriate echo displays.

The *display unit* is used for the presentation of the information combined in the video mixer. The most usual method of display for rotational radar installations is the plan position indicator (PPI). This method uses a c.r.t. with a circular screen in which the beam, during each sweep, is deflected toward the edge from the centre point. The rate of sweep is constant and is chosen to represent a particular range scale, which is usually adjustable in steps, and represents the slant range of any target on the radar screen. At the same time, provisions are taken to ensure that the direction in which the c.r.t. beam is deflected coincides with the actual azimuth of the radar antenna. The actual target display is achieved by initially making the c.r.t. beam perform its prescribed sweep pattern at a low illumination intensity. The video signal then causes a brightening of the beam with the result that the target appears as a spot of light, the brightness and length of which depend on the shape and amplitude of the received signal. The PPI display is basically a map presentation with features such as range rings and other marks representing geographically fixed objects which have not been removed by the MTI. Also it is possible to have a map-like presentation on which the targets are displayed in their relatively correct positions as measured in terms of azimuth and slant range.

There are many other types of display which are of no interest here since they are mainly used for special purposes.

Display units have various controls which allow the radar operator to select the range to be displayed, the brightness of the picture, and sometimes the antenna scanning rate.

After this brief description of a primary surveillance-radar unit, the basic function of a complete secondary radar interrogator station will be explained.

From the radar clock pulse generator the *coder* receives the secondary radar *pre-trigger* signal. Depending on the pulse-repetition frequency of the radar unit this trigger pulse is either accepted directly, or is first fed to a pulse divider circuit to produce a sub-harmonic. From this matched pre-trigger signal the interrogation signal is then shaped, in the video unit, in the form of two interrogation pulses $P_1$ and $P_3$ with an interval corresponding to the required interrogation mode. During this process it is possible to change modes after each interrogation in accordance with a prescribed pattern produced by internal or external programming.

In addition to the information solely restricted to the interrogation, the coder also supplies the following auxiliary signals.

(a) The SLS pulse, $P_2$ which is supplied to the modulator in common with the interrogator video signal.
(b) A control signal for the SLS switch.
(c) A mode gate for the decoder.
(d) A trigger pulse for the receiver gain time control (GTC-amplification control) which is dependent on range.

A receiver gate ensures that the receiver will only be 'on' during the radar period subsequent to the secondary radar interrogation. By this means the range can be unambiguously allotted to a radar echo and to the associated secondary radar response.

The output power to be processed in *the modulator* of the interrogator units is not a matter of primary importance. Particularly in the instance of modern transmitters, which use grid modulation, such a modulator can be achieved with semiconductor circuits.

However, this modulation signal, compared with the radar modulator, involves a complication connected with the fact that a secondary radar signal, in contrast with a primary radar transmitter signal, consists of a series of very irregular interrogation and SLS pulses between which no recovery time can be guaranteed. In addition, strict requirements are made regarding the shape of these pulses and their positioning in time.

In comparison with the transmitter in a primary radar unit the *transmitter* in the interrogation unit has a much reduced output power. To obtain the necessary frequency stability for a co-channel operation, it is quartz-crystal controlled. Thus, the initial stage is a quartz crystal which oscillates at a frequency of approximately 100 MHz and achieves the output frequency, conventionally, by several multiplier stages. Although the control unit in this transmitter is constructed with semiconductor components, the transmitter stage and its driver stages still use thermionic valves. In the installation chosen for our example the peak pulse power is 1.5 kW for a duty cycle $<1.5\%$. With this relatively low mean power it is possible to use (grid modulated) disc-seal triodes.

The diplexer is quite noticeably different from that in a radar unit. In the secondary radar method the transmitter and receiver frequencies are different and there is the further fact that there is a smaller difference in the signal levels during transmission and reception. Consequently, a dividing filter is used to isolate the transmission and reception paths.

The transmitter signal is only divided in the SLS switch. The actual interrogation pulses, $P_1$ and $P_3$ reach the directional antenna via the revolving turret, whilst the reference pulse $P_2$ is emitted by an omni-directional antenna.

There are several types of *secondary radar directional antenna*, as has already been indicated. In the example chosen, a linear array monopulse antenna, is used which is mounted on the upper edge of the radar reflector. The dimensions of this antenna, in the direction at right angles to the direction of propagation, are somewhat smaller than the dimensions of the radar antenna. For this reason, and also because the wavelength is longer than that of the primary radar, the horizontal beamwidth is wider; indeed, it is wider by a factor of approximately five. To compensate for this a monopulse technique is used to make an artificial reduction in the width of the displayed responses. The output signals from the two halves of the antenna are connected directly to a four-port network (ring hybrid) fitted to the antenna, which produces the sum ($\Sigma$) and difference ($\Delta$) signals. The two signals are connected to the receiver through separate channels in the rotating joint.

In the example under discussion the SLS antenna is fitted to its own mast. Whilst the horizontal diagram should be as circular as possible, the vertical diagram should be matched as closely as possible to the vertical diagram of the interrogator antenna. Here, very great care must be taken to ensure that the phase centres of the two antenna lie in the same plane above the ground, so that this coincidence should also be valid for the zero points or indentations produced by ground reflections.

The *receiver* has two almost identical channels so that it becomes possible to perform beam sharpening in one plane. Both receiver channels use a superheterodyne receiver and obtain their mixer voltages from the control transmitter. These mixer voltages are also derived (in the same phase) from the quartz crystal in the transmitter. The input sensitivity of the receiver (with a bandwidth of 8–10 MHz) amounts in this instance to $-82$ dBm, with a signal-to-noise ratio of approximately 10 dB. It is equally important that the receiver should also process its information without any distortion. To compensate for the difference in the field strength, when receiving responses transmitted over different ranges, there is included a GTC circuit for attenuating close-range responses. The behaviour of this circuit is similar to that of the STC circuit for attenuating close-range echoes in the radar unit. After rectification and regulation in the receiver the signal is supplied as video for further processing.

The purpose of the *defruiter*, which is a synchronous filter, is to suppress non-synchronous extraneous responses.

The transmitted information is processed in the *decoder* and is supplied in a form intelligible to the radar operator. Without explaining how they are engineered, an example of the three methods of decoding can be given:

(a) A passive decoder ascertains whether the incoming responses coincide with one of the expected codes. If this occurs, then a synthetic signal is sent at the correct time, via the video mixer, to the radar display so that the appropriate target can be marked.

(b) The response code train associated with a particular radar target that has been selected with the light pen is converted in the active decoder into numerical form, and is then displayed on a digital indicator.

(c) The information transmitted in response to a mode C interrogation is also decoded in the altitude decoder, where the code is converted in such a way that the aircraft altitude can be directly read off on a digital indicator unit in steps of 100 ft.

The *control units*, which control the interrogations, select the required type of decoding and set in the response code to be expected, are located near the display unit. They provide the link between the radar operator and the secondary radar installation and they facilitate the flow of information required for any specific operational task.

It is thus clear that although the primary and secondary radar systems are very much alike from the point of view of the location principle, yet, so as to transmit information, the secondary radar method requires numerous additional units, and the common display of information on the radar screen requires a very close linkage between the two installations. A relatively simple manual installation was described where the radar operator was part of the control loop in the flow of the information. This radar operator controls the SSR and questions the target, using the interrogator. After passing through the information channel the reply is processed and presented to the operator at the display in legible form.

Obviously, also, a computer can effect this evaluation process. But this only affects the interfaces for the input and output of the data. The rest of the installation remains largely the same.

The question arises regarding the specifications under which such installations are developed. In this respect it becomes evident that as regards specifications and transmission units, that is, the transmitter receiver and antenna installation generally have to be separately considered as regards the evaluation of information. Such a division is manifestly sensible, since the cooperation of the individual units with the whole secondary radar system is the critical factor determining the characteristics of the transmission units. On the other hand, the design of the information processing part is dependant on the basic idea, on the design and on the organization of the available air traffic control equipment.

## 2.1. Specifications for the Transmitter and Receiver Units in Secondary Radar Interrogation Units

The system specifications devised by ICAO in Annex 10 of the Communications Division are mandatory and must be observed by anyone wishing to operate a secondary radar system. This ensures that every interrogation station can operate any transponder, provided that this also satisfies the ICAO specifications.

Besides this outline specification, additional specifications have also been issued by the National administrative authorities; for example, the already mentioned *U.S. National Standard for Common System Component Characteristics* for the IFF Mark X-SIF/ATCRBS (Air Traffic Control Radar Beacon System), which controls the joint operation of civil and military secondary radar systems. Such national specifications can prevent other radio and navigational aids suffering interference due to the operation of secondary radar systems and vice versa.

Finally the specifications, as devised by the appropriate air traffic control services who operate these installations, must be considered as definitive for the detailed design of the equipment. As this concerns government bodies, who produce these specifications in close collaboration with their national radio authorities, these specifications acquire an official character.

Conversely, an important part of these specifications is devoted to the requirements regarding the use of military installations and to their environmental conditions. As regards maintenance and logistics these military units must generally satisfy much stricter conditions.

A typical specification for universally applicable secondary radar transmitter and receiver units, which is concerned with civilian air traffic control, is the FAA Specification R1212 of the U.S. Federal Aviation Authority. The fact that a vast amount of operational experience is available for units based on this specification has induced other countries to adopt it or, at least, to use it as an example.

To give an (abbreviated) idea of the design of these specifications a broad outline of the main divisions of specification FAA 1212 is given below (in fact some 100 pages are required to cover all the items).

*FAA 1212: Specification for a Secondary Radar Ground Station for Air Traffic Control*

1. Scope
2. The standards and specifications which are applicable
3. The requirements of the installation
3.1. Definitions:
   All the technical expressions and features specifically required for the units are here described and defined. In this section, for example, a detailed description is given of the procedure that must be used for making pulse measurements.

3.2. Equipment and service to be furnished by the contractor
3.3. Location of ground station equipment
3.4. Basic design requirements.
This section deals, for example, with the possibilities of connecting defruiters and the application of the SLS method. This section also lays down the amount of traffic that can be handled and the capacity for receiving overlapping responses.
3.5. Interrogation antennae and their accessories
3.6. Interrogation transmitter equipment characteristics
3.6.1. The transmitter
3.6.2. Interrogator coding units
3.6.3. The modulator
3.7. Receiver equipment characteristics
3.8. SLS – units
3.9. Not used
3.10. Video and trigger remoting equipment
3.11. Indicator site equipment.
This part of the Specification only covers, and provides a very brief description of, the minimum amount of equipment required for each decoder function. Provision is made, however, expressly for the possibility of connecting other decoding devices not described in detail.
3.12. Power supplies
3.13. Primary power
3.14. Standby channels
3.15. Control circuits
3.16. Measurements and methods of metering
3.17. Mechanical design
3.18. Test equipment, cables and accessories
3.19. General
3.20. ⎱ Modifications to other specifications superior to FAA specifications
3.21. ⎰ for their adaptation to the application mentioned in the present specification
3.22. Instruction books
3.23. Trouble-shooting manuals
4. Acceptance tests
5. Delivery preparations.

These specifications mention the minimum values of transmitter power and receiver sensitivity which must be attained to ensure maximum range performance. There is, however, an expressed requirement that it should be possible to intentionally reduce these values over a wide range. The intention behind this Specification is that the interrogation units should be adapted, as far as possible, to local conditions. When cooperating with a neighbouring radar unit it is not desirable that targets beyond its radar range should be interrogated by a secondary radar unit, since these cannot provide any useful information and merely magnify the interference inherent in the system such as fruit, overinterrogation, etc.

For a transponder which has to cooperate with any interrogator station, its characteristics are largely determined by the system and must be very strictly maintained. In the instance of interrogator units, however, there are a number of characteristics which can only be decided after designing a specific unit − sometimes only after making range and coverage measurements. The specification, consequently only gives the limits within which such adjustments may be made.

## 2.2. Concepts of Signal Evaluation

Since we have already established the criteria in accordance with which the radar observer can evaluate the information in a secondary radar response, two very different technical methods of realizing these criteria will now be discussed.

### 2.2.1. The common decoder

The common decoder (Fig. 2.2) is based on the simple, easily-observed units, used since the earliest days of secondary radar techniques. Each visual unit had a separate equipment for active and passive decoding.

With the increasing numbers of visual devices that began to be connected to radar units, an effort was made to find an economic solution to the idea of a single decoder. In fact passive and active decoders frequently perform similar functions which can be accommodated together in the common decoder.

*Spike Suppression*
To suppress narrow interference pulses the width of the incoming pulses is tested. Only these pulses with a width $\geq 0.3$ μs are passed on for further processing.

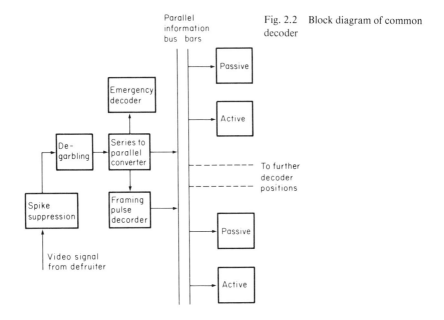

Fig. 2.2 Block diagram of common decoder

*Degarbling*
All replies are subjected to analysis to prevent the production of false information caused by overlapping pulse trains which could produce a false reply.

*Series Parallel Conversion*
The reply, which has been transmitted as a series of pulses in a temporal sequence (in serial form), must be available in parallel form for all the subsequent decoding processes, i.e. it must be available simultaneously at the required number of information lines. This series parallel conversion is usually performed by a delay line with tappings at 1.45 μs intervals. The parallel video signals are then processed via tapping amplifiers and impedance converters so that they can be connected to a large number of decoding devices.

*Decoding the Framing Pulses*
The first pulse in any aircraft response is made to coincide with the last pulse by means of the above-mentioned delay line. This coincidence signal indicates the location of an aircraft target responding in an SIF code. Within the decoder, this signal is used to read out the parallel information from the decoder delay line; outside the decoder this signal, indicating the

coincidence of the framing pulses, is used for special passive decoding displays (Beacon Assist, etc.).

The common decoder also contains the passive decoding units, which are only rarely required, for recognizing the various emergency signals (see Section 1.3) and I/P responses.

With this configuration, a considerably larger number of separate decoding devices can be connected to the common decoder.

As regards passive decoding, these units are essentially decoding matrices and code setting units.

With active decoding there must also be a stage for validating the data by comparison with the series of subsequent responses and then a stage for the conversion of the digital information into a numerical display.

The active altitude decoder and the active decoder are similarly constructed. It has, in addition, a number of code converters and facilities for including meteorological correction values in its calculations.

It is thus evident that the idea of a common decoder involves the concentration of several common circuit components in a central unit with a sort of bus-bar, providing the parallel video signals to which, as required, a fairly large number of active and passive decoders can be connected.

This concept is particularly advantageous where units of very different configurations are required within the same user organization.

In military applications it is usually a condition, for logistic reasons, that the same type of unit must be used in simple installations as well as in large complex systems. An inexpensive basic arrangement can then be extended, if necessary, at a corresponding cost.

### 2.2.2. Automatic decoders

In the instance of automatic decoding, which has already been described in Section 1.3.4, all the apparatus used is concentrated in a central location which is then sub-divided into two data-processing units. Only code setting and indicator units are required for the actual link between the radar display and the radar operator. The actual processes of active and passive decoding are achieved by the appropriate programming of the data-processing units.

In this concept of decoding even a basic lay-out is extremely expensive. However, additional equipment and further input and output units can be very easily added. Consequently, this concept is most suitable for large air traffic control and air space monitoring installations as must probably be obvious by analogy with presently available primary radar extractors.

## 2.3. Operational Reliability and Monitoring

Safety is naturally essential in aviation generally and this, obviously, is also true as regards air traffic control. Safety is dependent on the reliability of a large number of separate systems. The effect of the reliability of these individual units can have very different effects on the total reliability. In each instance this depends on the extent to which the functions of the faulty part of the systems can be promptly and sufficiently substituted by other systems.

If primary and secondary radar systems are to be used as the principal means in an air traffic control system, and if all the advantages of the minimum separation which they offer are to be fully used, then any return to the old methods of air traffic control to substitute for the failure of any radar sensor would lead to a very critical transition phase.

For this reason great efforts are made in radar systems and particularly in secondary radar systems to achieve very high system reliability. This can be achieved in a number of very different ways as the following paragraphs will show.

### 2.3.1. Civil air traffic control units

In secondary radar installations for civil air traffic control, the effort to maintain the high standards of reliability achieved, by careful development, of the individual units, is directed in the first instance to keeping the environmental conditions in the areas concerned as nearly constant as possible. For this purpose trained staff carry out maintenance services at short regular intervals during the course of which they have to operate an immense amount of metering and test equipment.

During maintenance operations, or if an installation should fail, a complete spare installation is always available. In air traffic control installations of this kind the redundancy principle is in operation whereby only a few installations, located at very carefully selected geographical positions, are available for the whole country.

Internal test circuits and special test transponders set up in fixed positions at a sufficient distance from the interrogator station are used to monitor these installations. These test transponders are generally developed in accordance with stricter specifications than the normal production runs of aircraft transponders. They are a form of metering equipment with which the functioning of the complete system and the maintenance of the essential tolerances can be continuously monitored. Any impermissible functional deviation can be detected so that an immediate automatic switchover can be made to the spare installation which is kept running in readiness. Alarm devices are used so that the maintenance services can give the appropriate assistance in the event of failure.

### 2.3.2. Military units

Military usage is entirely different from civil methods, where units in fixed locations are employed for air traffic control.

Many units are required for various military applications which must either be mobile, or where necessary must be moved from place to place or are a component part of a mobile weapon system. The units are, consequently, subjected to very severe environmental conditions, to high- and low-temperature conditions and to high shock and vibration stresses.

To develop reliable units for such conditions it is essential to subject the whole development project starting with the individual components, to regularly repeated environmental tests. Even when the units are being manufactured the costs of inspection and quality control are necessarily very high.

Another difference as compared with civil usage involves the staff structure. The actual user of a unit is usually only trained to carry out his own special task. Maintenance and testing cannot involve more than simple decisions as to whether a unit is still operational, or whether it should be sent to a field workshop for further examination. The best that one can expect of the operators is that, with the help of instruction manuals, they may be able to locate and replace a faulty sub-unit. To provide this sort of assistance military units generally have a considerable amount of built-in test equipment, the nature and number of these being dependent on the tactical purpose of the units. In the mobile military units in use today the basic amount of such equipment is the so-called BITE (Built-In Test Equipment), the principle of which is illustrated in Fig. 2.3.

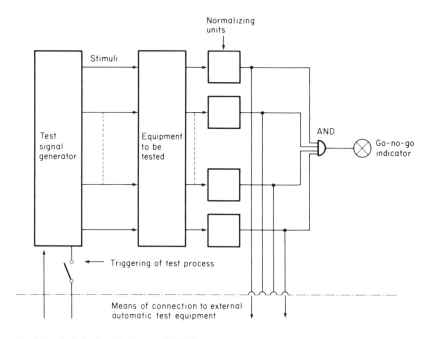

Fig. 2.3 Built In Test Equipment (BITE)

The unit to be tested has inputs to important points in the circuit to which the appropriate excitation signals, or stimuli, are applied by triggering the BITE test signal generator or, if necessary, by supplying a sequence of signals. The reactions of the unit to these stimuli are detected by means of sensors located at suitable points in the circuit so that the maintenance of the tolerances can be checked. By means of the right matching circuits each of the sensor signals is normalized and thus produces a 'yes' or 'no' decision regarding the proper functioning of the piece of circuitry under test.

Each of the standardized test signals is connected to a logical AND gate, so that a summary display 'unit in order' only appears if all the individual measurements are within the permissible tolerances. This decision by the BITE equipment is called a 'go-no-go' display.

In actual practice, these standardized sensor signals are not connected solely to the 'go-no-go' logic. They are also made available externally through a multiple plug connection. By means of suitable adapters a semi- or fully-automatic test programme can then be performed which will give a very reliable indication of the condition of the unit and can, if necessary, localize a fault.

If the number of signal inputs and test sensors is increased then, by varying the combinations of test signals and reactions, almost every fault can be detected by external automatic test equipment. Expensive automatic test equipment of this type is of course only used in the higher maintenance echelons.

By reason of the necessary mobility, the test transponders, usual in civil aviation, cannot generally be used so that in actual use the fault-free operation of the secondary radar unit can only be monitored by the BITE unit.

Spares in the event of a faulty unit are not possible in military operations owing to the lack of space in mobile units. It must be also realized that in their tactical role units can be put out of operation by enemy action. Under such circumstances any spares in the same location would also, in all probability, be destroyed. In military usage therefore every effort is made to scatter the units over a wide area and to monitor the coverage area with several independent secondary radar units, which will not usually be operating simultaneously.

## 2.4. Secondary Radar Interrogator Unit, Type 1990

During the design and development of the secondary radar interrogator unit, Type 1990* the military contractors had to make allowances for some special features. The interrogator unit, Type 1990, was planned as a universal equipment suitable for all the secondary radar applications then known to the German armed forces. However, even in the planning stage it became evident that a standard unit could not be made, because of the diverse requirements demanded of it, quite apart from economic considerations.

For this reason a modular system was developed whereby the complete interrogator unit is divided into sub-units which are individually interchangeable, and which are to a large extent independent of one another. The input and output values of these sub-units were chosen so that the spatial arrangement of the sub-units, in relation to one another, only has to

---

* The interrogator unit, Type 1990, the basic units for which had been developed by the Hazeltine Corporation, Green Lawn, New York, U.S.A., is manufactured in Germany by Siemens AG, Munich.

be limited in a few exceptional instances. The most suitable spatial arrangement for various types of usage can thus be arranged, in which the actual interchangeable parts, namely the sub-units, are always the same in any arrangement, the only differences being in the racks, the housing, chassis and the associated cabling.

When designing the unit the critical condition was its application as a mobile unit in confined spaces. The following stipulations were the result of this condition:

(a) The system had to be capable of extension into a more costly installation, starting from a minimum equipment consisting of a few basic units. This is the reason why the common decoder principle was used.

(b) Minimum volume and weight. This requirement has been satisfied by the extensive use of semiconductor components and integrated circuits.

(c) The environmental conditions are determined by the climatic requirements, that the equipment should be capable of unprotected operation out of doors, both in cold weather and in tropical areas and deserts. This means a temperature range of $-45°C$ to $+60°C$, in which the equipment must operate faultlessly. The shock and vibration resistance of the equipment must meet the conditions of the vehicle in which it is to be used.

(d) The power supply must be universal. The equipment can be connected to practically any single-phase supply voltage between 110 and 250 V, with a tolerance of $\pm 10\%$. The supply frequency may be either 45–65 Hz or 360–440 Hz.

(e) Particular importance is given to high operational reliability for use in remote stations, as, for example, on ships.

(f) The underlying principles of testing and maintenance must satisfy the military authorities.

The following text will describe and discuss in greater detail one of the possible arrangements of this equipment; namely, the rack Model 1990D which has been designed for use in military air traffic control. Figure 2.4 shows a front view of the rack when closed. The separate units are located in 19-in trays which can be pulled out on runners. Flexible leads are provided for the connections, so that maintenance is possible even during operation.

From the top, downward, the separate trays contain: (i) the transmission line connector unit for the video signal, the synchronizer and trigger pulse,

Fig. 2.4 Secondary radar interrogation unit, type 1990D

and the SLS switch with the control circuit; (ii) the oscillator, control transmitter, transmitter, modulator, two diplexers with two receiver mixers, two IF pre-amplifiers, the RF test unit and the transmitter power supply; (iii) two IF main amplifiers, test and monitoring equipment, and a power supply; (iv) the coder and common decoder (including the emergency decoder) and a two channel passive decoder; (v) the defruiter (a 'bought-in' unit); (vi) the multi-channel active/passive decoder.

In the following description of the Type 1990 secondary radar interrogation unit, the individual units will be treated in relation to the flow of information independently of any constructional relationship.

The overall block diagram (Fig. 2.5) shows, first of all, the possible extent of

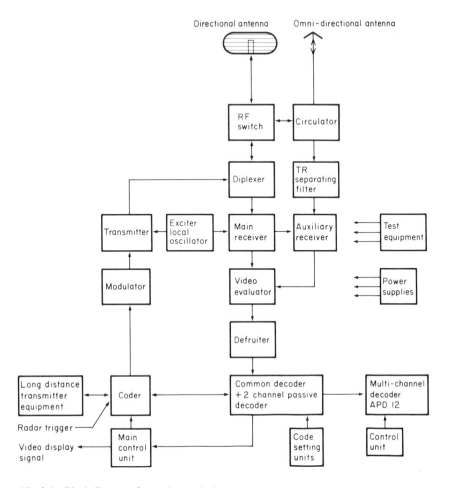

Fig. 2.5 Block diagram of secondary radar interrogation unit, type 1990D

an installation. The basic equipment consists of the transmitter, receiver, coder and common decoder. Defruiters, multi-channel active/passive decoders and remote control equipment are also available.

### 2.4.1. The coder

The coder (Fig. 2.6) is usually triggered from the primary radar by a pre-trigger signal with a lead time of 80–90 µs. If the radar pulse-repetition

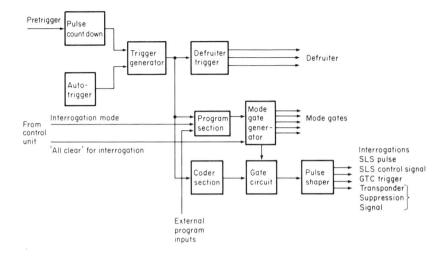

Fig. 2.6 Block diagram of coder in the interrogator unit, type 1990

frequency is above 450 Hz, then the pre-trigger signal (supplied at this same frequency) must be divided into a sub-harmonic below 450 Hz. A pulse divider circuit is available for this purpose. In the event of an emergency, for example, if the radar clock-pulse generator should fail, and also for maintenance, the coder is provided with an auto-triggering unit. Under such autonomous operation the trigger pulse required to trigger the radar display unit is produced at the correct time.

The coder produces the pair of interrogation pulses $P_1$ and $P_3$, for all civil and military modes by means of a delay line. The control pulse, $P_2$, which appears 2 μs after the first interrogation pulse (and which is required for the three-pulse SLS process) can either be mixed into the pair of interrogation pulses or can be supplied to a parallel output.

The following output signals are also produced:

(a) Mode gate pulses. These mode gate pulses store the interrogation mode until the return of the reply code and facilitate the collation of the interrogation and response information.
(b) A switching pulse to control the SLS switch (SLS Trigger).
(c) A pulse (GTC Trigger) to trigger the receiver amplification control, which is a function of time.

(d) A gating pulse to suppress any transponder associated with the interrogator station, i.e. to prevent its response to its own interrogations (transponder suppression gate).

(e) A pulse (defruiter trigger) to trigger the three-channel defruiters.

All the signals reach the coder output through pulse shaper stages and drive amplifiers. Externally-produced interrogation signals can also be applied for interface to automatic data processing devices.

The coder allows interrogations in different modes to occur in successive interrogation periods (mode interlace). The variation of interrogation modes is determined by a special programming unit which suits the operational conditions at the radar site in question. Its function is as follows:

(a) A cyclic variation of the interrogations in all the modes connected to the control unit.

(b) A limitation of the number of interrogation modes to 1, 2, 3 or 4 of the 6 possible modes so as to receive a sufficient number of hits from any aircraft that cannot reply in all the modes to all the interrogations. Because the situation may arise where the control unit has set a larger number of modes than can be used by the input program, the program is additionally provided with a series in descending order of priority within which the six modes can be arranged in any order. Within the group of one to four modes selected, the series of interrogations is then made in cyclic order.

(c) A limitation of the six possible interrogation modes to 1, 2, 3 or 4 with priority in selection is possible so that where more modes are used in the control unit than the maximum number available for the modes interrogation. If the control unit has fewer modes, then, after transmitting the interrogations in the modes available, no further interrogations are made for the number of periods required, until the cycle of 2, 3 or 4 possible modes starts once again. The result of this is a reduction in the interrogation duty cycle.

(d) If mode-interlace programs exceeding the limits of the 720 available arrangements are required, then an external programmer can be connected to the decoder. An output pulse to trigger the additional programmer and inputs to trigger the mode gate pulses are available.

### 2.4.2. The transmitter (Fig. 2.7)

The source of the production of a frequency-stable transmitter and receiver local oscillator frequency is the exciter local oscillator. For both instances

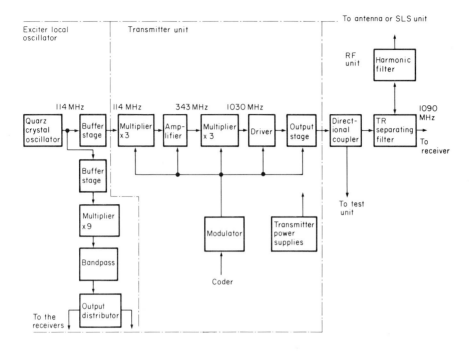

Fig. 2.7 Block diagram of transmitter interrogator unit, type 1990

the drive is a common quartz-crystal generator since the IF frequency of the superheterodyne receiver (60 MHz) is the exact difference between the transmitter frequency (1030 MHz) and the receiver frequency (1090 MHz).

Part of the output power of the quartz-crystal oscillator, which operates at 114.44 MHz is applied via an isolating (buffer) amplifier to a nine-times multiplier stage. The multiplier output is connected to a divider via a bandpass filter to ensure the required spurious frequency suppression. Both halves of the power supplied are connected to the main and auxiliary receivers where they are available as a local oscillator frequency.

Another portion of the output power from the quartz-crystal oscillator is connected from a second buffer/amplifier to a first multiplier stage and is then connected to the actual power transmitter through yet another buffer amplifier. The power transmitter consists of three identical coaxial circuits equipped with planar triodes. The first stage is a frequency multiplier, the two subsequent stages being the driver and the output amplifier.

The interrogation information is impressed on the transmitter signal by means of grid modulation. A transistorized modulator modulates:

(a) The 1st multiplier stage and the buffer amplifier.
(b) The 2nd multiplier stage, the driver and the output stage.

It should be noted that it is only these final stages which are equipped with valves: all the other parts use semiconductors.

The transmitter output signal is connected through a directional coupler to the preselector diplexer whence it is supplied to the 'main channel' antenna connector via a harmonic filter. The transmitter has its own special power pack to, principally, provide the (up to 2.6 kV) voltage for the valve anodes. By switching the low-voltage side of the power pack, the peak pulse power of the transmitter can be reduced, in several stages, from a nominal 1500 W down to 50 W.

### 2.4.3. SLS switching unit

In air traffic control installations the size of antenna chosen is large enough to provide a suitable beamwidth for display purposes. Consequently, in the unit described, no beam focusing by means of monopulse techniques is needed.

In this instance the auxiliary receiver channel is used for reply path side-lobe suppression (RSLS).

To suppress the side lobes on the interrogation path (ISLS) – where all the military units in the western world use the three-pulse method – in the Type 1990 equipment the $P_1$ and $P_3$ interrogation pulses are produced in a group, together with the SLS pulse $P_2$. The distribution of these pulses to the two antennae $P_1$ and $P_3$ to the directional antenna and $P_2$ to the omni-directional antenna, is achieved by means of a high speed RF switch. Figure 2.8 shows the block diagram for the SLS unit.

To be able to use a common omni-directional antenna for ISLS and RSLS, the main and auxiliary channels of the receiver are decoupled by means of a circulator. The actual RF switch uses *p-i-n* diodes as the switching elements. These are special semiconductor diodes which, at the carrier frequencies of 1030 MHz and 1090 MHz, lose their rectification properties. At these frequencies they simply behave as switched-in resistances which acquire a higher or lower value of resistance depending on their switched condition. It is also possible, through this inertial property, to permit the

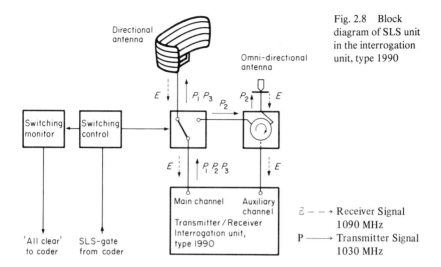

Fig. 2.8 Block diagram of SLS unit in the interrogation unit, type 1990

$E \dashrightarrow$ Receiver Signal 1090 MHz
$P \longrightarrow$ Transmitter Signal 1030 MHz

voltages and currents in the RF carrier oscillations to have far greater amplitudes than the switching voltages or currents would actually permit.

One of the problems in the selection of suitable *p-i-n* diodes is the required switching speed (governed by SSR system specifications) which must be very fast. Also, the diodes must be capable of withstanding the heat due to the transmission of the transmitter power over the range where the switch is closed and must be capable of withstanding the voltage in the range where the switch is open. The switching speed and power, however, in the production of *p-i-n* diodes are, within limits, mutually interchangeable. In the instance of a secondary radar system where the switching times must be less than 200 ns, the transmitter power more than 2 kW and the decoupling between the two switching outputs greater than 25 dB, each of the switching branches uses two *p-i-n* diodes (Fig. 2.9). The two *p-i-n* diodes in each branch are fed, in parallel, by means of a control circuit with a negative voltage in their 'off' (non-conducting) stage, or, with a positive voltage, over the conducting range of their d.c. characteristic. Apart from side effects they thus produce, for the carrier frequency, either an open-circuit or a short-circuit condition.

The diodes in the other branch are supplied by the control circuit with the appropriate complementary control quantities. Thus, if the diodes in branch 2 behave as an open-circuit, then the diodes in branch 1 behave as a short-circuit and vice versa. The appropriate values for the diode impedance, i.e.

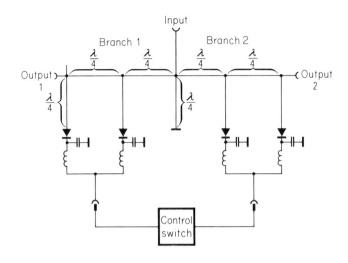

Fig. 2.9 Circuit diagram to show principle of the *p-i-n* diode controlled HF switch

Table 4 Characteristics of *p-i-n* diode switch

| | |
|---|---|
| Frequency range | $f = 1030–1090$ MHz |
| Transmittable power at a temperature of 85°C with matched outputs | |
|   Peak pulse power | $P_{max} = 4$ kW |
|   Mean power | $P = 22.5$ W |
| Transmittable power with short-circuited or open-circuited outputs | |
|   Peak pulse power | $P_{max} = 2$ kW |
| Permissible power during a failure of the control circuit at a temperature of 25°C | |
|   Peak pulse power | $P_{max} = 500$ W |
|   Mean power | $P = 10$ W |
| Attenuation in closed condition | $a_d = 0.5$ dB |
| Attenuation in open condition | $a_s = 40$ dB |
| Switching time | $t = 200$ ns |
| Drive per diode | ON $+50$ V  1 µA |
| | OFF $-100$ mA  1 V |

approximately a no-load or open-circuit condition, are changed by the quarter-wavelength conductor element at the branching point. A characteristic feature of this quarter-wave transformation is the fact that a short-circuit at the input appears as an open-circuit at the output, whence an open-circuit condition at the input is correspondingly transformed into a short-circuit at the output. By transforming the impedance values of the *p-i-n* diodes at the branching point, the switching can be performed without the need to have the appropriate components physically present at that point.

By combining the short-circuit and open-circuit conditions at the branching point, the signal path can be switched from the input to either of the two outputs of the *p-i-n* diode switch.

The *p-i-n* diode switch used in the interrogator, Type 1990 has the characteristics shown in Table 4.

The drive signals are produced by the SLS-gate pulse through a special switch drive-circuit. The control pulses are pre-distorted to obtain a high switching speed.

When the switch is operating normally, i.e. when the *p-i-n* diodes represent what is practically a short-circuit or open-circuit condition, the power loss in the switch is small. If, however, there is a fault in the diode bias voltages, then the diodes adopt an intermediate impedance value between the short- and open-circuit conditions. As a consequence, the effective power consumption is so great that, at the full transmitter power, the diodes can be destroyed. For this reason the control circuit and its output signals are monitored. Only when it is certain that the switch is definitely in one of its specified positions, is the interrogator control switched through to the transmitter via an 'all clear' signal in the decoder.

### 2.4.4. The receiver

To facilitate either the use of the monopulse method, for beam sharpening, or the suppression of side lobes in the reply path (RSLS) the receiver is of a two-channel type.

Both receiver channels are electrically identical (Fig. 2.10). The response signals received from the directional antenna are supplied to the main receiver channel. After passing through a harmonic filter and the diplexer, they are fed to the mixer stage. The diplexer is a frequency-dividing filter for separating the transmitter (1030 MHz) and the receiver frequency

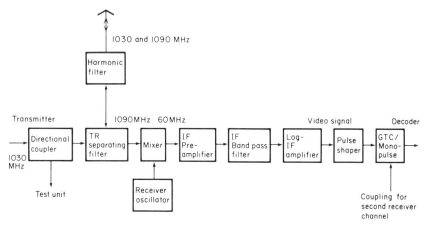

Fig. 2.10 Block diagram. One of the receiver channels in the interrogator unit, type 1990

(1090 MHz). It comprises coaxial cavities and is used to decouple the transmitter and receiver paths to preselect the receiver signals. This preselector also contains a special coupler for the input of a test signal from the testing and monitoring units. The mixer stage is located in the same unit as the diplexer.

By virtue of the heterodyne frequency from the control transmitter the receiver signal is converted to the 60 MHz IF frequency, and is then fed to the low-noise IF pre-amplifier. Because its output impedance is some 50 $\Omega$, the IF signal can now be sent over coaxial cable with hardly any practical limit to the distance from the receiver.

At the matched input of the IF post amplifier there is a five-stage band-pass filter, which basically determines the receiver selectivity. The amplifier has a logarithmic amplification characteristic and, consequently, has a very high dynamic range (>70 dB). It is designed on the successive detection principle and supplies the receiver signal at its output to the video stage. It is, of course, distorted due to the logarithmic effect in comparison with a signal from a linear amplifier. The video processing circuits, however (which are located in the IF unit) reproduce the original pulse form of the video signal.

### 2.4.5. The logarithmic IF amplifier and the processing of the signal

As is well known the response signal from a transponder may have very different levels at the receiver input. The fact that this level varies with range

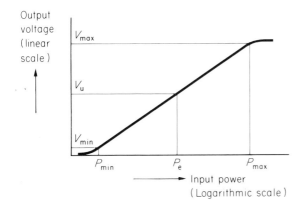

Fig. 2.11 *Transfer characteristic of a 'log-IF' amplifier*

in the proportion $1/R^2$, fadings in field strength, and shading produce differences of the order of 60 dB in the maximum and minimum signals to be processed. An input signal with such variations nevertheless requires further processing in logic circuits. It is, therefore, necessary to find a suitable method to achieve the transition from an analogue signal to a standardized digital signal. The standard methods of automatic receiver control are usually unsatisfactory with short response pulses, whilst limitations between rigid threshold levels may cause an impermissible increase in pulse width. Consequently, the method of 'amplitude compression' is used.

The logarithmic IF amplifier provides one means of achieving this. In such an amplifier the relationship between the input signal power $P_{r_i}$ and the output signal voltage, $V_a$, is

$$V_a = K \log_{10} \frac{P_{r_i}}{P_{r_i \min}}$$

This equation is only true within two limits determined by the technology of this device (Fig. 2.11). The upper limit is determined by the saturation, $P_{\max}$, of the output stage, the lower limit is set by the 'detection level' $P_{r_i \min}$ and a short transition area attached to it. Accordingly, the amplifier has a dynamic range $B$ (in dB) such that

$$B = 10 \log (P_{\max}/P_{\min})$$

There are several types of log-IF amplifiers:

(a) Amplifiers where the logarithmic compression is achieved in the separate amplifier stages in the IF stage (Fig. 2.12), because each separate, amplifier stage has a curved amplification characteristic.

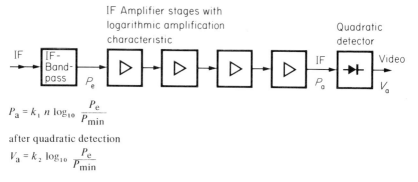

$P_a = k_1 \, n \log_{10} \dfrac{P_e}{P_{min}}$

after quadratic detection

$V_a = k_2 \log_{10} \dfrac{P_e}{P_{min}}$

Fig. 2.12 IF Amplifiers with logarithmic compression of the signal in the IF stage

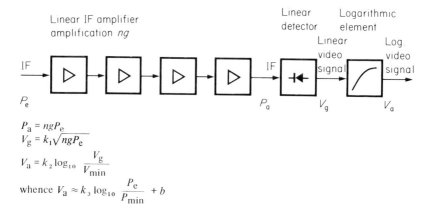

$P_a = ngP_e$
$V_g = k_1 \sqrt{ngP_e}$
$V_a = k_2 \log_{10} \dfrac{V_g}{V_{min}}$

whence $V_a \approx k_3 \log_{10} \dfrac{P_e}{P_{min}} + b$

Fig. 2.13 IF Amplifiers with logarithmic compression of the signal in the video stage

(b) Linear IF amplifiers, whereby the logarithmation is achieved, after linear rectification, in the video amplifier (Fig. 2.13). The difficulty with this method is that the IF amplification and rectification must have a linear characteristic over the whole dynamic range.

(c) IF amplifiers which operate on the 'successive detection' principle.

In this instance an approximately logarithmic characteristic is produced by successive detection at the separate amplifier stages of the IF amplifier (Fig. 2.14). In the first place, each amplifier stage has a constant amplification over its linear region, e.g. 10 dB, up to the point at which the amplifying transistor becomes saturated (Fig. 2.15(a)). In the second place,

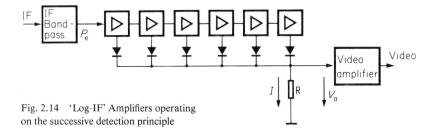

Fig. 2.14 'Log-IF' Amplifiers operating on the successive detection principle

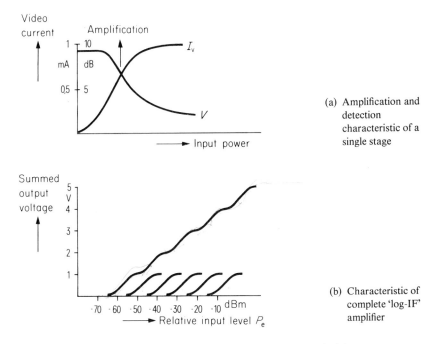

(a) Amplification and detection characteristic of a single stage

(b) Characteristic of complete 'log-IF' amplifier

Fig. 2.15 'Log-IF' amplifiers operating on the successive detection principle

the emitter/base junction voltage of the transistor has a threshold value. If the input signal exceeds the value then detection takes place. The current from the diodes flows via the buffer amplifiers to a summing resistance, where it produces that portion of the output voltage appropriate to each stage. By summing the individual characteristics in sequence, an approximately logarithmic characteristic is produced (Fig. 2.15(b)). (Note: If

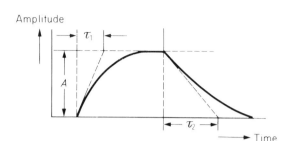

Fig. 2.16 Nomenclature for the individual quantities in a pulse with exponential leading and trailing edges

logarithmic amplifiers are built from stages with non-linear amplification characteristics, then all the selective devices must come before the input, otherwise there would be a considerable variation in frequency selection with any change in the input level.)

The next matter that has to be discussed is the signal pulse at the output of the 'log-IF' amplifier.

It is not helpful, in this instance, to keep to the usual description where the leading and trailing edges lie between 10% and 90% of the maximum amplitude. From the point of view of the dynamics of the receiver, this is only a fraction of the total pulse. A better understanding can be gained by assuming that the modulation signal has leading and trailing edges which are exponential functions (Fig. 2.16).

Thus, the leading edge becomes

$$V = A\left(1 - \exp\frac{-t}{\tau_1}\right),$$

the pulse top becomes

$$V = A$$

and the trailing edge becomes

$$V = A \exp\left(\frac{-t}{\tau_2}\right)$$

where $A$ is the amplitude; $\tau_1$ and $\tau_2$ are respectively the rise time and fall time constants.

These equations, describing the pulse form are substituted in the equation for the transfer function of the 'log-IF' amplifier. To do this, it will be useful to define a minimum input voltage $V_{r_i\text{min}}$, which corresponds to the minimum input power $P_{r_i\text{min}}$ at the input resistance $R_{r_i}$ of the amplifier.

By doing this, at the amplifier output the pulse top becomes

$$v_a = K \log_{10} \frac{V_{r_i}^2}{R_{r_i}} \frac{R_{r_i}}{V_{r_i\text{min}}^2}$$

or

$$v_a = 2K \log_{10} \frac{V_{r_i}}{V_{r_i\text{min}}}$$

or

$$v_a = 2K \log_{10} \frac{A}{V_{r_i\text{min}}}$$

the leading edge becomes

$$v_a = 2K \log_{10} \frac{A}{V_{r_i\text{min}}} + 2K \log_{10} \left(1 - \exp \frac{-t}{\tau_1}\right)$$

and the trailing edge becomes

$$v_a = 2K \log_{10} \frac{A}{V_{r_i\text{min}}} + 2K \log_{10} \exp \frac{-t}{\tau_2}$$

or after transformation (when using logarithms to the base 10)

$$v_a = 2K \log_{10} \frac{A}{V_{r_i\text{min}}} - 2K\, 0.43 \frac{t}{\tau_2}$$

or

$$v_a = 2K \log_{10} \frac{A}{V_{r_i\text{min}}} - 0.86\, K \frac{t}{\tau_2}$$

The first summand in these expressions gives the video amplitude for the top of the pulse.

From a comparison of the output pulse (Fig. 2.17) with a number of input pulses of similar shape but different amplitudes, it becomes clear that this involves merely a vertical shift in the vertical direction. For the pulse width, which in the linear presentation of pulse form was given between the points of half amplitude, another definition is widely used, namely the pulse width at the half-power level. This pulse width can be measured at the video pulse output at a voltage level which always lies below the pulse top by a fixed

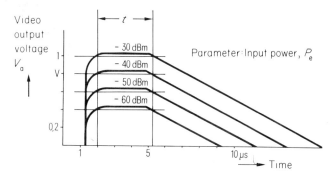

Fig. 2.17 The video output signal for pulses with the same exponential leading and trailing edges but with varying input power

voltage value. This fixed voltage value can be found by calculation since
$$\Delta V = -0.301\, K$$
where $K$ once again is the proportionality factor of the 'log-IF' amplifier.

Armed with this knowledge there is no difficulty in developing a circuit to regenerate the video signal (Fig. 2.18). For this purpose the video pulse is delayed, in a delay line, for a period $T$ (usually a fraction of the pulse

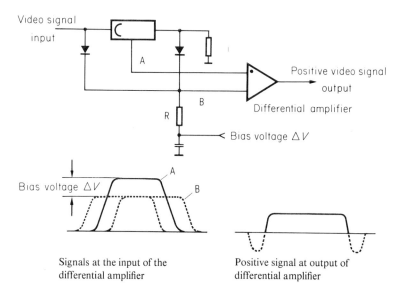

Fig. 2.18 Regeneration of a video pulse after logarithmic distortion

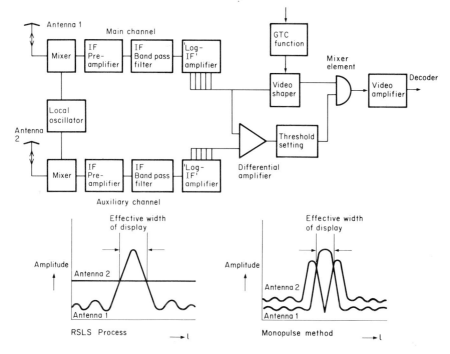

*This is true if the main and auxiliary channels have the same amplification

Fig. 2.19  Two channel receiver. Possible evaluation circuits for RSLS and monopulse methods

width), and is then fed together with the undelayed video pulse into an OR gate consisting of two diodes and a resistor. At the output of this gate will appear a pulse B, extended in width by the time $T$, from which a constant voltage $\Delta V$ is subtracted. This pulse B can thus serve as the reference level for measuring the pulse width. The video signal, with its original amplitude, is taken from a tapping on the same delay line giving a delay of $T/2$, and is then compared with the reference pulse, B, in a comparator. In this way it is only positive differences that are formed, with the result that the output of the differential amplifier supplies a regenerated video pulse which once again corresponds with the original pulse in width and amplitude.

At the video level a comparison of the signals from the main and auxiliary receivers is also made so as to decide whether to sharpen the beam by the monopulse type of method or whether to suppress the side lobes in the receiver path (to perform RSLS) as shown in Fig. 2.19.

One of the necessary conditions for this is that the signal level in the main channel should exceed that of the auxiliary channel by a specified amount. This operation can also be performed in a differential amplifier with a limitation on one of its inputs. A threshold value circuit is also provided to prevent interference due to noise being evaluated in the absence of proper signals.

The 'log-IF' amplifier has one special advantage; namely, that control of sensitivity can easily be achieved. For this purpose it is only necessary to subtract a voltage from the video signal, which will either correspond to the GTC time function or will also correspond with any adjustment due to manual control. With multi-channel receivers this method is particularly advantageous since such control only takes effect after the monopulse or RSLS evaluation and, consequently, makes a substantial simplification in the synchronization of the amplification characteristics of the main and auxiliary channels.

### 2.4.6. The defruiter

The defruiter used in the interrogator Type 1990, Configuration D, is a digital defruiter, Type DD200A (Standard Elektrik Lorenz AG, Stuttgart, Zuffenhausen) (Fig. 2.20) in which integrated circuits are used for the storage elements.

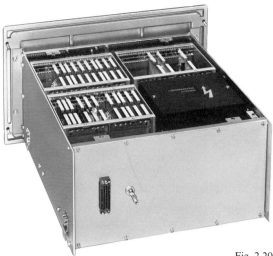

Fig. 2.20   Defruiter, type DD 200A

Upon receipt of the mode trigger signal, and a start information signal, the receiver video signal is advanced into the store by an independent timing arrangement, not synchronized with the interrogation timing, at a frequency of 1.33 MHz. This timing frequency produces time increments of 0.75 μs corresponding to range increments of 112 m. Because the defruiter timing is derived from a free-running oscillator ($f = 31.33$ MHz) by means of a 16:1 frequency divider, this produces an internal 'jitter' of 47 ns.

The defruiter DD200A has been designed for a maximum range of 120 nautical miles. This corresponds to a storage capacity of 1974 bits. In the immediately following radar dead time, the information cannot stay still, since a dynamic shift register is used. The information therefore has to be advanced through 26 further bits of the shift register, and thus bridge the dead time by advancing it at a slower timing which also corresponds with the minimum clock frequency of the shift register.

By means of the defruiter Type DD200A it is possible to effect a mode interlace interrogation with three different modes. Moreover, three different evaluation criteria can be set in, namely, '2 out of 2', '2 out of 3' and '3 out of 3'. For the range of 120 nautical miles, mentioned above, the total storage capacity of the shift register must be 12 000 bits. Metal oxide semiconductor (MOS) ICs are used. Each shift register, 100 bits in length, is housed in a TO 5-size 'transistor casing'.

Since the range of coverage is basically only a matter of the length of the register, it becomes simple to produce other types of defruiters for maximum ranges of 60, 180 and 240 nautical miles merely by omitting or adding storage units.

In addition to the range increments an important quantity for evaluating the efficiency of a defruiter is the width of the acceptance gate. The acceptance gate produced by the comparison of the delayed video from the initial interrogation period, with the video signal from the immediately subsequent period is 1.15 μs in width, if the information pulse coincides exactly with the range increment. If an information pulse spans two neighbouring storage cells, the width of the acceptance gate so produced is 1.9 μs. From these two values the resultant mean effective width of the acceptance gate becomes 1.6 μs.

On connection to the Type 1990 interrogator unit the defruiter receives the GTC trigger, as its start pulse, from the coder in addition to the mode trigger signals which are applied to separate inputs. The video signals are taken to and from the receiver to the decoder, via coaxial cables.

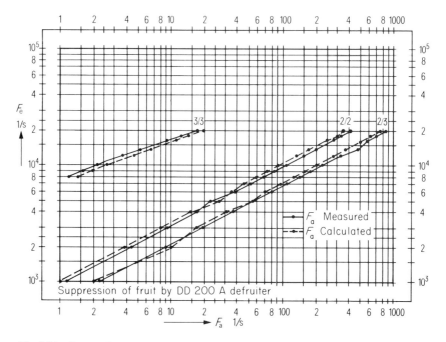

Fig. 2.21 Suppression of fruit by DD 200A defruiter

The diagram in Fig. 2.21 gives some information on the fruit suppression capabilities of the Type DD200A defruiter. Figure 2.22 indicates its acceptance characteristic.

### 2.4.7. The decoder tray

In the basic equipment of the Type 1990 installation the decoder tray contains a common decoder and a two-channel passive decoder.

The response signal from the receiver comes first to the decoder input circuit, which contains the pulse shaping circuits and a pulse width discriminator. The signal then passes through two delay lines, each with a length of 20.3 μs. The first of these delay lines is provided principally for the degarbling circuit whilst the second is used for series to parallel conversion.

To explain how a degarbling circuit operates, the example of the unit shown in Fig. 2.23 will be used. It has already been mentioned that the purpose of

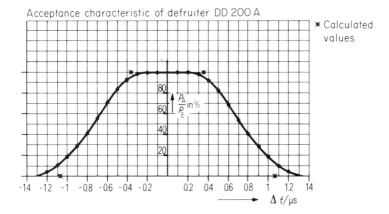

Fig. 2.22  Acceptance characteristic of defruiter DD 200A

this circuit is to prevent the occurrence of false information due to synchronous overlapping when reply pulse trains run into one another. If, on the other hand, the overlapping of responses is not synchronous, such that the pulse positions of the one train fall into the space positions of the other then the information in both trains is preserved and can be read out.

For this purpose the presence of a response or frame-pulse coincidence signal is first ascertained in an AND gate, 1, from the first delay line DL1. A further 20.3 µs frame-pulse coincidence signal is formed by the

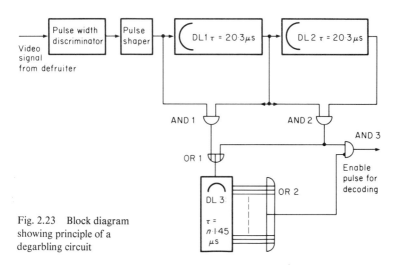

Fig. 2.23  Block diagram showing principle of a degarbling circuit

subsequent decoder delay line DL2 in an AND gate, 2. The two frame-pulse coincidence signals are combined in the OR gate, 1, and are fed to the delay line, DL3. This delay line, with tappings at 1.45 µs intervals, also has a length of 20.3 µs. Each output from the delay line, DL3, is connected to the multiple input OR gate, 2. In this way a sequence of time slots is formed which occur at the 14 raster positions, at intervals of 1.45 µs before and after each frame-pulse coincidence signal for the AND gate, 2.

Only under the condition where no frame-pulse coincidence signals coincide with one of the time slots can there be a guarantee that no synchronous garbling is taking place. In this instance the frame pulse coincidence signals from DL2 pass through the AND gate, AND 3, which has one inverted input, to provide the decoder with the 'enable' signal for reading the parallel video signal.

An exception here is the combination of an SPI, or emergency repetition signal, with a C2 pulse, which simulates an instance of garbling. A special circuit is provided which, in this instance, disconnects the garbling suppressor circuit.

Because the decoder also contains a two-channel passive decoder, a circuit illustrating the function of this unit is also shown.

By definition, passive decoding is a means of ascertaining whether the pulse pattern of a returning (reply) pulse train coincides with the pattern which is to be expected from the interrogation and which has been selected at the code-setting unit. When this happens, the appropriate radar target is marked on the face of the display with an additional pulse. To perform this function the reply video signal which enters in serial form (the pulses following one another in time) is first converted into parallel form. As Fig. 2.24 indicates this is performed by a delay line, DL2, with a length of 20.3 µs, which has tappings at intervals of 1.45 µs. In a similar manner to the degarbling circuit, a framing pulse-coincidence signal is produced from the input and output signals from the delay line. This signal indicates the exact moment in time at which a sequence of signals with the 'standardized' framing pulse interval characteristic of secondary radar responses is standing in the delay line.

When garbling of pulse trains is not detected the framing pulse-coincidence signal is passed via the appropriate suppressor unit to locations where it is required in the system. In some cases it is used directly, for example for the display of 'all aircraft' or 'beacon assist' information. For these applications the pulse is reshaped in monostable stage MK1.

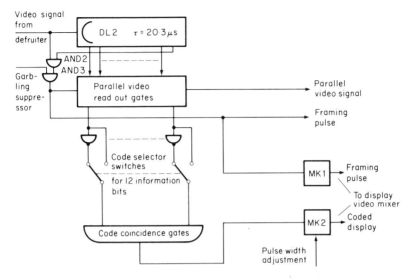

The code selector switches are actually electronic switches operated by the appropriate mode gate

Fig. 2.24   Block schematic showing principle of passive decoding

The fact that the time for reading the delay line and, consequently, for passive decoding, is determined by the coincidence pulse is a matter of greater importance. The tappings on the delay line are connected via buffer amplifiers and pulse shapers to a decoder matrix. This is an arrangement of logical switching units in which the incoming pulse train is compared with the code selected by the code setting unit. Only if there is complete coincidence in all the pulse (raster) positions is the monostable stage MK2 also triggered, via the AND gate, in conjunction with the frame-pulse coincidence signal. This MK2 unit produces the pulse required for the marking on the face of the display. The pulse length of this display pulse can be externally adjusted at the monostable stage. It therefore becomes possible to produce different thicknesses for the bars on the radar screen to suit different types of passive decoding displays.

With passive decoding special devices to check the data are not required, since correctly decoded answers are integrated in the phosphor of the radar screen to produce the marker display. Single incorrect pulse trains, in passive decoding, only involve the absence of the decoded signal, and, therefore, merely produce a gap in the series of correctly decoded signals. This gap, however, is usually beyond the resolving power of the radar screen and therefore never becomes evident.

The above explanation has been simplified to render it more easily understood. In actual fact the design of the decoder matrices has to be much more complex to ensure that all pulse trains are decoded, even where the pulse positions deviate from the ideal position, provided that they are within the permissible tolerance. Moreover, such pulses as exceed these tolerance limits may not be accepted so that the pulse train must be rejected.

The comparison of the reply pulse trains and the settings of the code selector unit with the appropriate interrogation mode is performed by first energizing the setting up leads via the appropriate interrogation mode gate.

The decoder tray in the 1990 installation has another component, a fixed emergency call decoder. This is a passive decoder which has a fixed wire code setting for each standard emergency call. The most important types of emergency call are the following:

| | |
|---|---|
| Code 7700 | to interrogations in modes 3/A and B |
| Military emergency call with the framing pulses repeated thrice | to interrogations in modes 1, 2 and 3 |
| Code 7600 | to interrogations in modes 3/A and B. |

If any of these emergency calls appear it is immediately recognized but may not immediately produce an alarm. To avoid false alarms a counter circuit is used to ascertain whether several alarm calls are arriving from the same target during a single sweep of the antenna. The optical and acoustic alarms are only operated if this number exceeds a specified number set to satisfy the prevailing environmental conditions. An alarm unit, similar in design to the other control units, is provided at the working position to initiate the sounding of the alarm and to stop it.

The processing of the parallel video signal for the other decoder devices is a further important function of the decoder. A special output amplifier is provided for each parallel bit so as to provide sufficient power and to achieve the required matching to the coaxial lines.

At this point it is necessary to examine the control components for the coding and decoding functions already described. These are located in two separate control units, the main control unit and the code setting unit.

The main control unit (Fig. 2.25) contains, basically, the setting mechanisms for controlling the interrogation, the display of the information and the receiver sensitivity as well as the associated indicator lamps. To

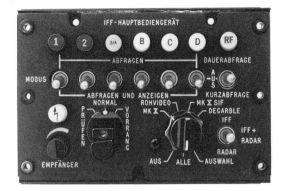

Fig. 2.25 Main control unit, type 1992

*Note:*
The switch position 'Select' produces the 'Beacon Assist' function

select the interrogation sequence a toggle switch is allotted to each mode. By means of another switch the interrogations can be triggered as continuous or momentary interrogations, whereupon the coder programming device determines which of the interrogation modes set in will actually be used. Used modes are shown by illumination of mode indicator lamps. Another indicator lamp illuminates if the interrogations are emitted at a sufficient level.

Two rotary switches enable the following variations to be set in for the passive decoding displays.

The *first rotary switch* has the following positions:

(a) Raw video display
(b) Only basic Mark X replies
(c) Mark X-SIF-replies, where the passive decoding is correct, and
(d) Mark X-SIF-replies, where the passive decoding is correct, with the addition of degarbling.

A *second rotary switch* makes it possible to switch in two further functions: namely 'All aircraft' whereby all responses to a specified mode are displayed jointly, independently of the mode set in; and 'Beacon assist' whereby all aircraft replying to the interrogation mode are displayed, with special preference for those aircraft replying in the correct code.

The selection and mixing of the primary and secondary radar signals to be displayed is also carried out in the main control unit. A switch is provided which enables primary radar data, secondary radar data or both simultaneously, to be displayed.

Fig. 2.26 Code setting unit, type 1993

Selection of codes for different modes is carried out by means of thumb-wheel switches on the code setting unit (Fig. 2.26). Typically, individual codes or groups of codes can be selected in one or more modes. The thumb-wheel switches are coded in binary form. Activation and selection of the code setting switches is controlled by the mode gates.

The main control unit and the code setting unit can be located up to 100 m from the main rack. Multi-core cables are used to connect the units together. If more than two displays are required to be connected to a single code setting unit an auxiliary amplifier is used. It is also necessary to provide a special unit which allocates priority to one display at a time.

Fig. 2.27 Multi-channel passive and active decoder APD 12

### 2.4.8. The multi-channel decoder

The two passive decoder channels in the common decoder are not sufficient when using the Type 1990 interrogator unit for air traffic control, because normally each of several aircraft using different codes usually have to be simultaneously observed at each operating position. For this reason an additional multi-channel decoder, of the APD 12 Type (Fig. 2.27) is provided in the D racks. It is housed, as an auxiliary unit, in its own tray and is designed to be connected to the common decoder. A complete unit contains two six-channel passive decoders and a single two-channel active decoder (Fig. 2.28). By removing some of the sub-units, the passive decoder can be converted into a single six-channel passive decoder.

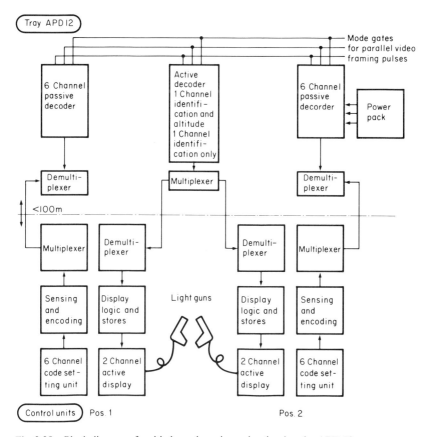

Fig. 2.28 Block diagram of multi-channel passive and active decoder APD 12

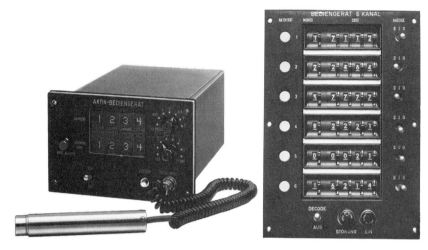

Fig. 2.29   Control units for APD 12

The multi-channel decoder, APD 12, can be simultaneously controlled by two display units where each display unit has a single six-channel passive control unit and a single two-channel active control unit (Fig. 2.29). Each radar observer thus has six passive channels allotted to him. The active decoder, which is only available to them one at a time, is shared by both observers by means of time division multiplexing.

The four control units can also be located up to 100 m from the rack. They contain all the necessary indicator and control components as well as some sub-units for the associated electronics.

The passive decoder channels are equipped with decoder matrices in the same way as the two passive channels in the common decoder. Two special features must, however, be mentioned.

*Multiplex Transmission between the Control Unit and the Decoder*
To set up the decoder matrices a minimum of 12 connecting lines is required for each code that can be set up.

In a six-channel control unit, therefore, 72 cores in a multiple cable must be provided solely for setting up the codes. Under military operating conditions this is a disadvantage, firstly because the breakage of a wire or similar failure could cause an incorrect code to be set into the decoder matrix and, secondly, because there are no simple means of recognizing and indicating such a fault.

For these reasons, 'serial transmission' of the code information is the method adopted and a single coaxial cable is used. This cable can be monitored easily and, also, provides effective protection from stray interference signals. In contrast with the serial transmission of secondary radar responses from the aircraft to the interrogator unit, the data for setting up the codes can be transmitted between the control unit and the actual decoder at a uniform velocity via a cable free from interference. The design of the data transmitter and data receiver – in this instance, therefore, on principle also a coder and decoder – can be very much simpler than that of a secondary radar coder and decoder.

*Activity display*
The purpose of a multi-channel passive decoder is, first of all, to set in a series of expected codes and then, if necessary, to decode and display these simultaneously. The radar observer is thus faced with the difficulty that he only has a limited supply of markers to distinguish the separate codes on the display screen, usually only 'off', a bar ('Normal') and the double stroke ('Twice'). In practice, therefore, the radar observer sets in codes from six expected targets and usually marks these with the 'Normal' symbol. If he now requires to know which mark belongs to which code, he switches the separate codes in succession to 'Off' or to 'Twice', and discovers, by the change in the display, which code belongs to which display. During this process he must also switch in those code settings for which at that instant no replies come in. To avoid this waste of time an activity indicator has been introduced which lights an indicator lamp adjacent to each code setting to indicate whether a reply in this code has been received in the course of the last revolution of the antenna. The radar observer can thus see at a glance which codes are in use at any moment.

Figure 2.30 gives a block diagram of the passive decoder unit in the APD 12 with its associated control unit.

Before stating the characteristics, the operation of the active decoder will be explained by means of the functional block diagram (Fig. 2.31).

In the active decoding process a selected reply pulse train has to be decoded and then displayed in digital form. To ensure that only the marked pulse train will, in fact pass on for further processing, the parallel video signal is first fed to an acceptance gate which is energized by the target selector unit (a light pen or area gating control). A mode gate connected between the target selector and the acceptance gate then ensures that only replies to the appropriate interrogation modes are decoded.

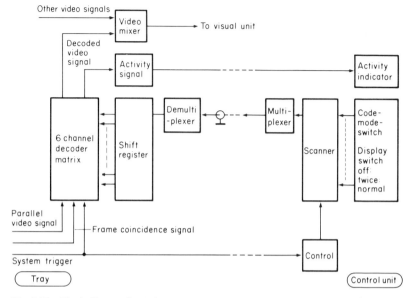

Fig. 2.30 Block diagram for 6 channel passive decoder APD 12

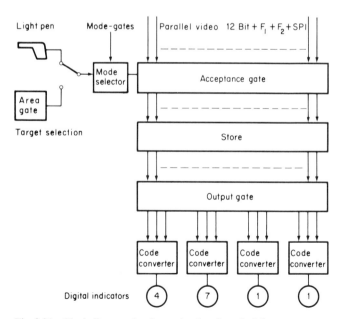

Fig. 2.31 Block diagram showing active decoder principle

In contrast with passive decoding, during the active decoding process a single falsified pulse train immediately causes the display of a false code. For this reason the parallel video signal from the acceptance gate is supplied to one or several registers for storage so that it can subsequently be compared at later intervals with replies appropriate to the same interrogation mode. Only when this check on the data shows that the stored information has a high probability of being correct – and this is based on specified safety criteria – is the parallel video signal passed, via an output gate, in four groups of three bits to the binary-to-octal code converter in the digital indicator. This digital indication then remains on display until the active decoder again receives a fresh target selector instruction whereby fresh reply information is received.

In the APD 12, the active decoder has two channels, the first of which, depending on the setting of the associated 'mode' switch either displays identification (for modes 1, 2, 3/A, B, D) or shows flying altitude (for mode C). The second channel is designed for identifications (Modes 1, 2, 3/A, B and D) only. Where altitude has to be ascertained in the first channel a special code converter has to be connected between the parallel video store and the digital indicator, to convert the MoA Gilham altitude code, as recommended by ICAO, into a decimal code (see Section 1.2.3).

To take account of local meteorological conditions there is a facility on the APD-12 tray for selecting a QNH correction within the range $-7900$ to $+7900$ ft. Where this correction has to be taken into consideration it is switched on by the control unit.

The altitude is indicated with the sign bit in 100 ft steps from $-1000$ up to $+126\,700$. In this display the last figure is separated by a comma from the three figures in the thousands positions.

The active decoder is connected to two active decoder control units located at the operating positions. It is only available to one control unit at a time but even if both operating positions should almost simultaneously trigger a decoder process a control switch ensures access to both positions without any noticeable delay. For this purpose, it is also necessary to locate the binary to decimal converters for the display units within the control units and to make these converters store the displays. In a similar manner to the passive decoder section of the APD 12 the connections between the active decoders and the control units are made by means of serial transmissions via coaxial cables.

In the APD 12 each control unit is fitted with a light pen as the basic equipment for target selection. It is also possible to have inputs from gated signals under the control of an 'area' gate.

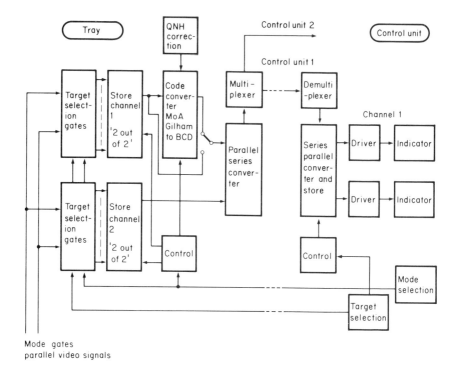

Fig. 2.32 Block diagram of active decoder in APD 12

The criterion for the accuracy of the data was, intentionally, made as simple as possible. A reply is therefore deemed to be probably correct if, within a sequence of seven samples, two consequent replies are identical. Figure 2.32 shows the block diagram for the active decoder section in the APD 12 with the associated control and display units.

### 2.4.9. Remoting equipment

The Type 1990 D interrogator unit, as described earlier, is intended for applications where the antenna, radar equipment, interrogator unit, and display units are close together. The maximum permitted cable distance between the interrogation unit and the display units is 100 m. If greater distances are required the cost of the interconnecting cable and the line

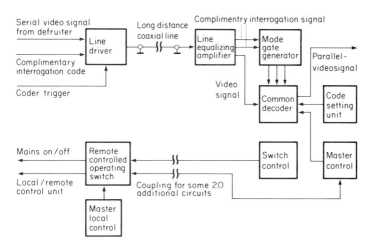

Fig. 2.33 Block schematic. Connection of remote operating positions to interrogator unit, type 1990, at ranges between 100 m and 3 km

attenuation have to be taken into account. At distances approaching 100 m it is more economic, as in the case of the APD 12, to combine the signals from the code selector switches and digital indicators in data transmission equipment.

Similar considerations arise when the interrogator equipment is used for civil air traffic control. An operational facility may be located on the same site as the secondary radar, but in addition there may be other operational facilities at remote locations. Remote distances up to 3 km are common in these circumstances.

Figure 2.33 shows how this problem was solved for the Type 1990 D units.

By means of a remote-controlled 'control' switch the interrogator unit ascertains the location from which the installation is being operated – i.e. by which location the selection of the interrogation modes and the triggering of the interrogations is being effected. The location which is not selected can only receive the interrogation modes then in use. When this switch is set to 'Remote' the mains supply can be switched-on from the remote location.

The reply pulse train is still transmitted to the remote location in serial form through a coaxial cable from the output of the defruiter via a line input amplifier, LIA.

127

In addition to the reply information, the interrogation information is also required for evaluation. Consequently a mode signal is transmitted via the same coaxial cable. In this instance this signal is a pair of pulses with an interval that is complementary to the interval of the interrogation pulse. Because there is a time interval between the interrogation and the response the two groups of signals can be transmitted on the same line.

Since a time reference signal is required the coder trigger is also transmitted along the line. It is, for discrimination, transmitted at a substantially higher amplitude.

At the other end of the coaxial line there is a line-equalizing amplifier, LEA. In the first place this unit adjusts the signal level by inserting a larger or smaller number of attenuators according to the length of the line. After this the pulse distortions caused by the coaxial cable are compensated in a delay compensator. This can also be adjusted in steps to the required line-length. After these measures, however, the level of the transmitted signal is so reduced that amplification is necessary. A wideband amplifier (0–15 MHz) is used for this purpose. Thereafter, the code trigger is extracted by an amplitude filter, and the interrogation signal and reply pulse train are separated from one another by means of time cut-outs. These can be connected to the mode gate generator via three separate lines. Although the coder dealt with here is a common decoder, of the type already described, it also contains additional equipment to decode the transmitted mode signals and the mode gates produced from these mode signals. All the other components in the unit, such as the control units, the Type APD 12 multiple decoders, etc. can be connected to the mode gate generator, in exactly the same way as they are connected to the common decoder in the Type 1990 interrogator unit.

### 2.4.10. Characteristics of the secondary radar interrogation unit, Type 1990

The essential characteristics of the secondary radar interrogation unit, Type 1990, are once again presented in a brief and compactly arranged form in Table 5.

Table 5  Electrical Characteristics of the Secondary Radar Interrogation Unit, Type 1990

| | |
|---|---|
| *Transmitter* | |
| Frequency (Quartz crystal stabilization) | 1030 MHz |
| Tolerance | ±200 kHz |
| Modes | 1, 2, 3/A, B, C and D |
| Adjustment of interval between the control Pulse, $P_2$ and the first interrogator Pulse, $P_1$, for side lobe suppression on the interrogation path (if mixed with the pair of interrogation pulses) | 1.7 µs–2.3 µs |
| Transmitter power (peak pulse power) (may be reduced to) | 1500 W (min) 500, 150 and 50 W |
| *Receiver* (The data are applicable to each of the two receiver channels) Frequency | 1090 MHz |
| Absolute tolerance of the local oscillation frequency | ±200 kHz |
| Intermediate frequency | 60 MHz |
| Video output Adjustment (limited) of output voltage Impedance | 1–5 V 75 Ω |
| Sensitivity (for 50% correct decoding of framing pulses) | −80 dBm |
| Dynamic range | 60 dB |
| GTC (Gain Time Control) Range of regulation (7–30 µs after trigger pulse) As a result of the amplification adjustment the amplification increases by 6 dB every time the range is double until 'normal amplification' is reached. The maximum deviation for this variation in amplification is ±3 dB | 10–35 dB |
| 6 dB Bandwidth | 8–11 MHz |
| Two-channel operation for RSLS or monopulse evaluation | |

*Coder*
Mode interlace  up to 6 modes
(Adjustable internally to a max. of 1, 2, 3 or
4 modes)

Defruiter trigger pulses for 3 channel defruiter  3

External programme inputs for any sequence of modes  6

Synchronization
With primary radar system: Pre-trigger
triggering pulse period prior to radar transmitter
pulse  80–90 µs

Range of adjustment of interrogation pulse
frequency by dividing the internal (adjustable)
frequency self-triggering pulse from the pre-trigger  150–450 Hz
ISLS pulse, $P_2$ (can be obtained either at a separate
output or in combination with the pair of
interrogation pulses)

*Decoder*
A common decoder containing two passive decoding
circuits such that each circuit can recognize each of the
codes selected in all the modes.
Video mixer permitting the choice of:
(1) Basic Mark X*; Raw video Mark X-SIF, or degarbled video
(2) Beacon assist.
Number of parallel outputs  15
Degarbling and decoding of framing pulses
Emergency recognition
 Basic Mark X – emergency
 Mark X-SIF – emergency
 Code 7700 in interrogation mode 3/A and B
 Code 7300 in interrogation mode 1
 Code 7600 in interrogation mode 3/A or B
SPI recognition

Electromagnetic compatibility as per military
specification  MIL-I-26600

*Power Supplies*
Power consumption  Approx. 400 VA
Voltage (can be selected)  110, 117, 127, 150
  190, 204, 220, 234,
  or 250 V $\pm$ 10%
Frequency (can be selected)  45–65 Hz or
  360–440 Hz

* An older military identification system.

*Design Data for D Type Rack*

The sub-units are housed in 19-in rack trays.

The dimensions of the rack are:

| | |
|---|---|
| Height | 1430 mm (max.) |
| Width | 601 mm |
| Depth | 637 mm (including tray handles) |

The weight of a fully equipped rack is approximately 250 kg.

The control units and the alarm unit are contained in a control unit housing.

The interrogator unit assembly has been designed to satisfy the following (military) specifications:

(a) Drip-proofing as per MIL-STD-108
(b) Vibration and shock resistance as per MIL-T-4807

These basically involve

Vibration tests at the resonance points, Method 1A, 5–20 Hz. Deflection 1.5 mm. Vibration resistance testing as per Method 1A 15–55 Hz, deflection 0.38 mm, for a period of 45 min. in each of the three axes.

(c) Requirements generally expected from electrical equipment for use in cold climates and under tropical and desert conditions as per MIL-E-4138.
(d) Requirements for testing aviation equipment under environmental conditions as per MIL-E-5272C with a temperature limitation to the range $-45°C$ to $+60°C$.
(e) Other requirements regarding humidity, salty spray, reduced pressure, and usage on vehicles.

# 3. The Secondary Radar Transponder

The purpose of the secondary radar transponder, as has already been explained, is to receive interrogations from a secondary radar interrogator station, to recognize these interrogations and to transmit a coded reply.

As the block diagram, Fig. 3.1, shows, the interrogation signal is received by an omni-directional antenna which is connected to a diplexer and then routed to the receiver. This is usually a superheterodyne receiver in which the local oscillator is stabilized by a quartz crystal. After demodulation the output signal from the receiver is supplied to the decoder as a video signal. The decoder ascertains whether the interrogation presented to it is of standard form and performs a test to see if the interrogator signal is free from the characteristics of side-lobe suppression, ISLS. If this check is positive it then ascertains the type of interrogation beamed to the transponder. The result of this decoding is supplied as a mode gate to the coder. The coder thereupon compiles a reply pulse train which not only corresponds to the mode gate, and, therefore, to the interrogation, but also agrees with the code setting which the pilot has set on the control unit. A further input is provided for the transmission of altitude information in reply to a mode C interrogation. In this instance the coder is connected to the encoding mechanism on the pressure altimeter system.

The output signal from the coder – a video signal – is input to the modulator which modulates the reply transmitter, creating the response.

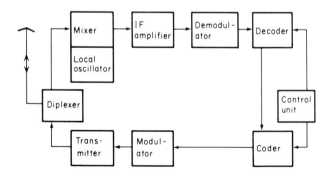

Fig. 3.1   Secondary radar response unit (Transponder)

The response signal is routed by the diplexer to the common omnidirectional antenna, which radiates the response back to the interrogator station.

## 3.1. Specifications for Secondary Radar Transponders

At this point it would appear appropriate to gain some knowledge of the specifications for SSR transponders.

Although they are actually dealing with a single system, there are differences in the format of the specifications for interrogators and transponders. Obviously, one requirement common to both is that every interrogator station should be able to communicate the appropriately arranged data with every transponder over the required transmission distance.

However, there is a very great difference in the number of equipments available. Compared with the small number of interrogator units there are a very large number of transponder units. In 1969 there were already more than 200 000 private aircraft in the U.S.A., the greater number of which were equipped with an SSR transponder. In such a situation some balance has to be struck between the technical effort and economic burdens involved in ground and aircraft equipments. From the size of the potential market it can readily be understood that there is keen competition amongst a large number of aircraft instrument manufacturers, as regards their SSR transponders, which is generally reflected in the price level. The owner of a small private aircraft may see no immediately perceptible benefit from the incorporation of an SSR transponder and, consequently, chooses the most inexpensively manufactured item if he finally decides to purchase such a piece of equipment at all.

In a situation of this sort the licensing authorities in each country must raise certain minimum requirements for aircraft instruments. Using the U.S.A. once again as an example, Fig. 3.2 shows the relationships between the various specifications.

In the U.S.A., the FAA is empowered by law to issue regulations for the properly-conducted operation of air traffic control services and, therefore, for the operation of secondary radar systems. These require, from the point of view of the system, that the international ICAO regulations in Annex 10 should be observed. It is a matter of importance that the user of these regulations should give equal consideration to the interests of commercial

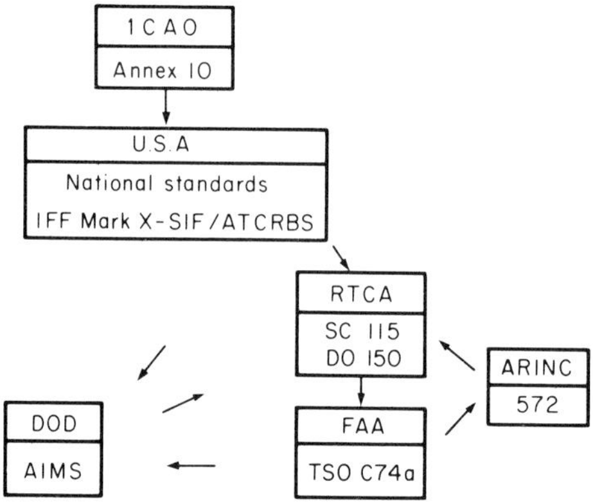

Fig. 3.2 U.S.A. specifications for the secondary radar system

```
        ┌─────────────┐
        │    ICAO     │
        │  Annex 10   │
        └─────────────┘
        ┌─────────────────────┐
        │        U.S.A        │
        │  National standards │
        │ IFF Mark X-SIF/ATCRBS│
        └─────────────────────┘
                ┌──────┐
                │ RTCA │
                │ SC 115│
                │ DO 150│      ┌───────┐
                └──────┘       │ ARINC │
                               │  572  │
        ┌─────┐  ┌──────┐      └───────┘
        │ DOD │  │ FAA  │
        │ AIMS│  │TSO C74a│
        └─────┘  └──────┘
```

aviation, to privately-owned aircraft and aircraft used by businesses without forgetting considerations of national defence. Moreover, any interference with other air traffic control and radio services must be avoided. These *National Standards* have already been mentioned in the specifications for interrogator units. They are applicable to every section of the secondary radar system.

An intermediate stage has been included for the preparation of SSR transponder specifications, namely *The Radio Technical Commission for Aeronautics (RTCA)*. Representatives of the authorities, industry, aviation companies and relevant associations work together in this commission. Within RTCA, the Sub-Committee SC 115 is responsible for SSR transponders. This Sub-Committee has devised a minimum performance standard (MPS), denoted DO 150, which takes proper account of the interests of the authorities, industry and the users.

A special body which works in cooperation with RTCA must be mentioned here, namely Airlines Radio Incorporated (ARINC). ARINC is a development firm founded by the U.S. airways companies, and maintained by them to compile common technical/engineering principles, to draft regulations and specifications, and to test instruments as regards their suitability for use on aircraft. Thus, for example, ARINC has also published a special specification for the SSR transponder, ARINC 572. It is a customer specification, the customer in this instance being the Association of U.S. Airlines.

On the basis of the recommendations of RTCA, the FAA aviation authorities in the U.S.A. are now issuing minimum performance standard

Table 6  An Extract of the Essential Characteristics of Specification TSO C74a

| | |
|---|---|
| Receiver frequency | $1030 \pm 0.2$ MHz |
| Selectivity at $\pm 25$ MHz | 60 dB |
| Sensitivity | $-72$ to $-80$ dBm |
| Spurious sensitivity | 60 dB |
| Interrogation modes | Minimum 3/A and C with the pairs of pulses as per ICAO Annex 10 |
| SLS<br>Decoder tolerances | Acceptance $\pm 0.2$ µs<br>Rejection $\pm 1$ µs |
| 'Grey Zone' | 9 dB |
| SLS dead time | $35 \pm 10$ µs |
| Pulse width discriminator<br>Echo suppression | See Section 3.2.3 |
| Dead time | 125 µs |
| Adjustable overload limit | from 500–2000 replies s$^{-1}$ |
| Reply codes | 4096 + SPI |
| Transmitter frequency | $1090 \pm 3$ MHz |
| Transmitter power<br>(Peak pulse power) Class I<br><br>Class II | 250 ... 1000 W with up to 1200 replies/s with complete code 7777 & SPI<br>100 ... 600 W with up to 1000 replies/s with complete code 7777 & SPI |
| Period between interrogation and response | $3 \pm 0.5$ µs |
| Jitter | $\pm 0.1$ µs |
| Reply coding | As per ICAO Annex 10 |
| Altitude code | MoA-Gilham code |
| Antenna | Omni-directional pattern with vertical polarization $\pm 30°$ with respect to the horizontal (not possible in practice!) |
| Environmental conditions | Depending on category of usage |

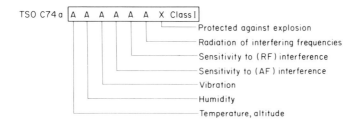

Fig. 3.3 Key to conditions for employment of equipment according to TSO-specifications

specifications for aircraft instruments. For the SSR transponder this specification is TSO C74a (TSO, Technical Service Order) the essential content of which is shown in Table 6.

In the event of different categories of use, TSO C74a makes provision for possible variations regarding particular points in the specification. This is indicated by adding a group of letters in addition to the statement of the TSO number (Fig. 3.3) which indicate the type of environmental conditions to be expected where each of the letters has the following significance.

*First Letter*
Altitude and temperature.

| Category | A | B | C | D | E | F |
|---|---|---|---|---|---|---|
| Maximum operating altitude in ft | 45 000 | 30 000 | 20 000 | 15 000 | 20 000 | 12 000 |
| Operating temperature at sea level in °C – from – to | −54 +55 | −46 +55 | −40 +55 | −15 +55 | −15 +40 | −15 +40 |

*Second Letter*
Requirement regarding humidity

Category A. Normal
Category B. Subject to high humidity for fairly long periods.

*Third Letter*
Mechanical stress determined at the installation position.

| Description of aircraft | Installed in | | |
|---|---|---|---|
| | Tail | Instrument panel or anti-vibration rack chassis | Elsewhere |
| Helicopter | A | F | A |
| Turbo driven | B | C | A |
| Multi-motor, piston engined | E | F | D |
| Single motor, piston engined | F | F | F |

For categories A to F there are also detailed tables which give numerical values for the stresses to be expected at the locations mentioned.

*Fourth Letter*
Sensitivity to AF interference.

| Category A | For units installed in aircraft where the a.c. generator installation is of greater power than 250 VA |
|---|---|
| Category B | For units installed in aircraft which have no a.c. generator or have such an installation with a power equal to or less than 250 VA |

*Fifth Letter*
Sensitivity to RF interference.

| Category A | Equipment for installation in aircraft with a total weight greater than 5600 kg |
|---|---|
| Category B | Equipment for installation in aircraft with a total weight less than 5600 kg |

*Sixth Letter*
Spurious radiation.

Categories A and B as for the categories applicable to the fifth letter.

*Seventh Letter*
Resistance to explosions.

| Category X | For all normal installation conditions where special anti-explosion precautions are not necessary |
|---|---|
| Category E | For installation in locations where an explosive environment must be taken into consideration |

This group of aphabetical letters follows data on another classification for equipment.

Class I: Aircraft for use at altitudes of 10 000 ft or higher.
Class II: Aircraft flying at a maximum altitude of 15 000 ft.

The data regarding the difference in the transmitter power and mean reply rates for Classes I and II is not at first very intelligible from a systematic point of view. The following points, however, must be taken into consideration.

(a) Class II transponders are normally used in very small aircraft. In such aircraft the antenna leads are shorter and consequently cause less attenuation. The value given for this is therefore 1.5 dB.
(b) For Class II transponders, the maximum flying altitude of 15 000 ft corresponds, at the wave length specified, to a radar horizon of approximately 280 km, that is therefore approximately 75% of the total range for the system.

In conjunction with the specification work done by the civil authorities the Ministry of Defence, DOD (Department of Defence) in the U.S.A. produces another set of specifications under a programme entitled AIMS; although these are 'tailored' for military use, they are also based on ICAO Annex X. In these specifications special consideration has to be given to much stricter environmental conditions, to precautions against wilful interference by an enemy, and to the security of information regarding identification.

After describing the organizations for specifications in the U.S.A. some reference must now be made to the fact that Europe has also formed an association similar to RTCA, called *The European Organization for Civil Aviation Electronics* (EUROCAE), whose head office is in Paris. A special working committee WG9 deals with MPS (Minimum Performance Standards) for SSR transponders wherein, in distinction to the U.S. specifications, the special geographical, political and organizational conditions appertaining to Europe are taken into consideration.

The remainder of this chapter will discuss a representative section of available transponders. Here it is advisable to divide transponders into two groups according to engineering effort and cost: IFF transponders on military aircraft, and SRR transponders for air traffic control.

## 3.2. IFF Transponder AN/APX-46

The AN/APX-46 equipment will be discussed as the first example of a military IFF transponder. This transponder was developed in 1958 by the Hazeltine Corporation, Greenlawn, New York, U.S.A., and was manufactured in Europe by Siemens AG, Munich. With some adjustments it conforms to the new AIMS system specification issued in 1964. Thousands of units in all the NATO states have successfully proved the reliability of this equipment. The specifications are listed in Table 7. At the time of its development, its transistorized design was a complete novelty and is still of interest today.

In the following paragraphs more recent units will be mentioned but will not be explained in detail.

### 3.2.1. Design and block diagram

With the IFF Transponder AN/APX-46 it is possible to design special housings, for the various units, which will be best suited for use in different types of aircraft. In the Siemens AG programme, there are four different configurations (Fig. 3.4) the external appearance of which is very different, but which, basically, only involve differences in the housings and cable runs. By this means it is possible to simplify storage, to standardize staff training procedures, to unify test and maintenance procedures and lastly, but not the least important, to have economic conditions of manufacture with the increase in serial production.

The block diagram (Fig. 3.5) of the IFF transponder AN/APX-46 corresponds largely with the concept already outlined above. For design reasons some of the functional units are located together in one unit and some are split up into sub-units.

Table 7  Characteristics of IFF Transponder, Type AN/APX-46, After Modification to AIMS Specifications

| | |
|---|---|
| Transmitter power | $P = 500$ W (250 ... 1000 W) |
| Duty cycle | 1% |
| Transmitter frequency | $f = 1090 \pm 10$ MHz detunable |
| Stability of transmitter frequency | $f = \pm 3$ MHz under all operating conditions |
| Transmitter pulses<br>  Width<br>  Rise time<br>  Fall time | $T = 0.5 \pm 0.1$ μs<br>$t_1 \leqslant 0.1$ μs<br>$t_2 \leqslant 0.2$ μs |
| Receiver sensitivity | $P = -75$ dB normal<br>$-65$ dB low |
| Receiver frequency | $f = 1030 \pm 10$ MHz adjustable |
| Local oscillator | quartz crystal controlled  $f = \pm 200$ kHz |
| Dynamic range of receiver (when decoding) | 45 dB |
| Dead time | 125 μs |
| AOC | 3000 replies<br>  responses $s^{-1}$ for Mode 1<br>1500 replies<br>  responses $s^{-1}$ for Mode 2 |
| Code system | SIF (Selective Identification Feature) |
| Decoder | Mode 1: $3 \pm 0.2$ μs<br>Mode 2: $5 \pm 0.2$ μs<br>Mode 3: $8 \pm 0.2$ μs<br>Test mode $6.5 \pm 0.2$ μs*<br>Mode C: $21 \pm 0.2$ μs |
| Coder | Mode 1: 5 information pulses<br>Mode 2: 12 information pulses<br>Mode 3: 12 information pulses<br>Mode C: MoA-Gilham code.<br>The information pulses contained between two framing pulses spaced $20.3 \pm 0.1$ μs apart, occur at time intervals which are multiples of $1.45 \pm 0.05$ μs |

\* A special test interrogation in the test mode can be made in military IFF transponders as a test before take off, without interfering with any other operation. The reply is returned in the code selected by Mode 3.

| | |
|---|---|
| Identification | I/P identification of position by repeating reply after an interval of 4.35 µs |
| Emergency | Emergency code by repeating the reply framing pulses three times at intervals of 4.35 µs |
| Side-lobe suppression | 3 Pulse method |
| SLS criterion | $P_1P_2 = 2 \pm 0.2$ µs |
| 'Grey Zone' | 9 dB |
| SLS dead time | $T_s = 35 \pm 10$ µs |
| Auxiliary signals | Input suppression pulse<br>Output suppression pulse |
| Power supplies | 115 V/400 Hz, approximately 100 VA 28 V d.c. for tripping relays and illumination |

*According to MIL-E-5400 Class 2*

| | |
|---|---|
| Ambient temperature | $-54°$ to $+71°C$ (at normal air pressure) up to $+95°C$ intermittent |
| Vibration | Without shock absorbers 2 g at 0 ... 750 Hz<br>With shock absorbers 10 g at 0 ... 750 Hz |
| Shock | 15 g (11 ms) |
| Maximum flying altitude | 100 000 ft |

Weight and volume are dependent on the arrangement of the system.

*Typical data for RT 555/APX Arrangement*

| | |
|---|---|
| Weight | 12.3 kg |
| Volume | 15.7 l including vibration mounting |
| Blower incorporated | |
| Test unit incorporated | |
| Shock absorbers incorporated | |

Fig. 3.4 Four different configurations of the IFF transponder AN/APX-46

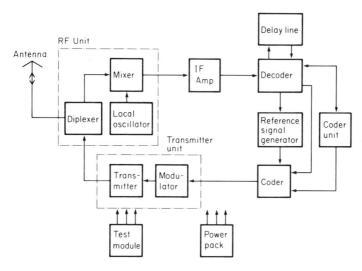

Fig. 3.5    Block diagram of IFF transponder AN/APX-46

### 3.2.2. The RF unit

The RF unit comprises the diplexer, the mixer and the local oscillator (Fig. 3.6). In this unit the diplexer for isolating the transmitter and receiver paths deserves some mention. As the sectional view and equivalent circuit (Fig. 3.7) show, it consists of a cavity unit by which the line from the transmitter to the antenna behaves substantially as a through-line without junction points. The filter section in the direction next to the mixer output has a bandpass characteristic with a mean transmission frequency of 1030 MHz and therefore blocks the transmitter frequency. This therefore ensures that the transmitter power reaches the antenna socket.

Conversely, precautions have to be taken to ensure that the interrogation signal reaches the receiver section. In the classical separating filter a bandpass filter for the 1090 MHz transmitter frequency would have been provided for the purpose in the transmitter section. In such a bandpass filter, however, there will be resonance rises which − if the mechanical dimensions are to be kept within sensible limits − at great flying altitude and, consequently, at low atmospheric pressure, can cause voltage breakdowns. For this reason the line from the transmitter to the antenna must be made as nearly as possible a constant through-line. Thus, for

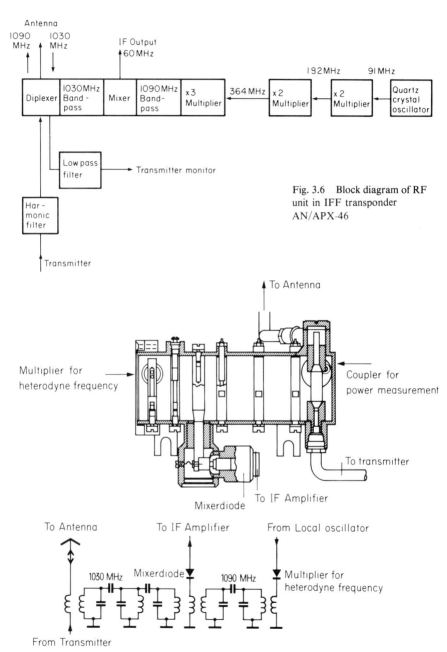

Fig. 3.6 Block diagram of RF unit in IFF transponder AN/APX-46

Fig. 3.7 Receiver separator in IFF transponder AN/APX-46

decoupling, the length of the line between the separating filter and the transmitter is so chosen that a receiver signal at a frequency of 1030 MHz will see the transmitter from the branching point, by reason of the line transformation, as an open-circuit.

The receiver signal thus reaches the mixer diode via the triple-circuit bandpass filter. The superheterodyne frequency required for the conversion is derived from a quartz-crystal oscillator and is then increased in a series of multipliers by the factors $2 \times 2 \times 3$. The tripler diode for the last multiplication is actually housed in the diplexer (preselector). This frequency then reaches the mixer diode via a double-circuit bandpass filter which ensures that the heterodyne frequency is satisfactorily free of spurious frequencies. From the mixer the IF signal passes, via a matching transformer, to the IF amplifier by means of a coaxial plug connection.

### 3.2.3. The IF amplifier unit

In addition to the IF amplifier, the IF Amplifier Unit also contains the demodulator, equipment for setting the level for the minimum receiver signal, a pulse-width discriminator for the suppression of spikes, and the echo suppression circuit (Fig. 3.8).

The actual IF amplifier is a four-stage transistor amplifier with a centre frequency of 59.5 MHz. The first three stages are cascode circuits. The cascode circuit, as Fig. 3.9 shows, consists of two transistors. The input transistor is operated as an emitter-follower, and the output transistor operates in a common-base circuit. In the present special IF amplifier, the

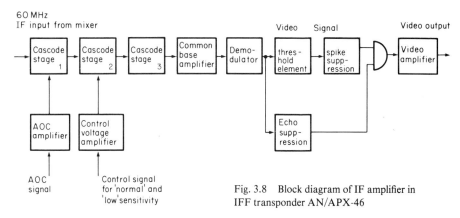

Fig. 3.8 Block diagram of IF amplifier in IFF transponder AN/APX-46

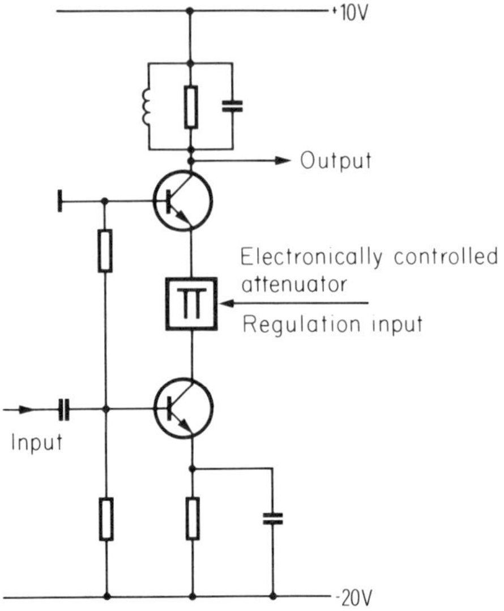

Fig. 3.9 Circuit for cascode amplifier

two transistors in the cascode stage are electrically coupled via a variable attenuator.

The two stages of a cascode amplifier are highly mismatched. Consequently, the power gain of a cascode stage is also less than the product of the gains of the two separate transistors. The cascode stage, however, has the following advantages:

(a) The input resistance is high, and corresponds approximately to the input resistance of an emitter follower stage.
(b) The output resistance is also high, corresponding approximately to the output resistance of a stage in a common base circuit. The high input and output resistance simplify the coupling to the frequency determining circuits between the separate cascode stages.
(c) In a cascode circuit the feedback from output to input is very low.

The fourth IF amplifier stage operates in the grounded-base configuration. The output circuits of the four amplifier stages have staggered tuning to produce a bandwidth of 8 MHz.

The transponder contains an automatic overload control (AOC) (see Section 1.9.2). The reference signal generator first determines whether there is an overload. If this is so, an AOC signal is supplied to the first cascode stage with the result that its amplification will be reduced to such an extent

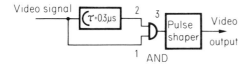
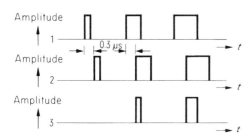

Fig. 3.10 Block diagram showing principle of spike suppression with pulse timing diagram

that only a number of interrogations exceeds the input threshold so that the act of responding to them is within the permissible load of the transmitter.

The second cascode stage has an input for setting the total amplification to 'Normal' and 'Low'.

A signal limiter is provided in the third cascode stage to prevent the subsequent stage from being overdriven.

The pulse-width discriminator for spike suppression (Fig. 3.10) can be very easily understood. The signal arrives at a delay line in which the delay of 0.3 µs corresponds to the smallest pulse which needs to be accepted. By connecting the input and output signals from the delay line to an AND gate, an output pulse will only be obtained from the gate if the input pulse is wider than 0.3 µs. The output pulse is, of course, reduced by 0.3 µs compared with the original pulse but it can be regenerated again in the subsequent video amplifier stage.

To explain the echo-suppression circuit it is worthwhile looking at the relevant regulation in the ICAO specification entitled *Echo Suppression and Recovery*.

'The transponder shall contain an echo-suppression facility designed to permit normal operation in the presence of echoes of signals in space'. By 'echoes' is meant such interrogations as reach the transponder by some indirect route, and which are therefore somewhat delayed and (usually) also less strong. The echo suppression must operate in harmony with the devices for side-lobe suppression.

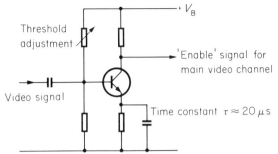

Fig. 3.11 Echo suppression circuit in IFF transponder AN/APX-46

In particular, it is specified that, after receiving a pulse with a minimum width of 0.7 µs and an amplitude $A$ above the receiver threshold, the receiver sensitivity is to be reduced. The range of the sensitivity reduction lies between $(A-9\text{ dB})$ and $A$. Within the first microsecond after starting echo suppression an overshoot is actually permitted on the reduction of the sensitivity beyond $A$. Directly after each suppression the sensitivity shall recover at a rate of 3.5 dB µs$^{-1}$.

This is achieved in the actual circuit by connecting the signal into a parallel branch via an echo-suppression amplifier, a threshold amplifier with an $RC$ section (with a time constant of 20 µs) in its emitter circuit (Fig. 3.11). Upon the termination of any pulse the charge on the capacitor is maintained with the result that the threshold for the next pulse is reduced approximately by the amplitude of the preceding pulse. As the length of time increases, the displacement of the threshold value decreases exponentially by the discharge of the capacitor. An output pulse is thus only produced if it has a specific amplitude dependent on the time and amplitude of the preceding pulse.

Thus, in the instance of side-lobe suppression, SLS, a decision is also made as to whether the amplitude relationship between the first interrogation pulse, $P_1$, and the SLS pulse $P_2$ satisfies the criteria for main-lobe and side-lobe interrogations as described in Section 1.6.1 and in Fig. 1.32.

The echo-suppression output pulse will be used, after amplification, to switch the original pulse through to the output via an AND gate.

### 3.2.4. The decoder unit

The decoder unit (Fig. 3.12) contains all the logical elements in the decoder. The required delay line is a separate, hermetically-sealed sub-unit. The

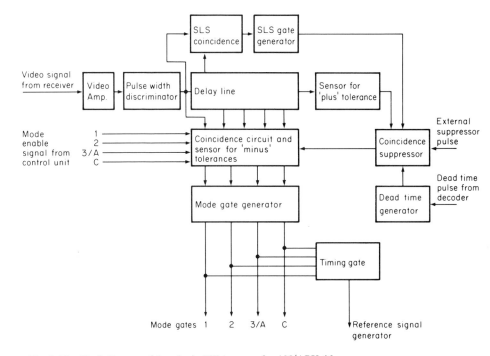

Fig. 3.12  Block diagram of decoder in IFF transponder AN/APX-46

interrogator signal, after passing through the video amplifier, is shaped into pulses of constant width, and thus reaches the delay line.

The first possible signal sequence to be considered in the SSR system is the pair of pulses with a 2 μs interval, characteristic of a side-lobe interrogation. To decode this 2 μs pair of pulses there is a 2 μs tapping on the delay line. The signal at this tapping is compared in an AND gate for coincidence with the input signal. Where the interval between pulses is not exactly 2 μs, nevertheless a coincidence pulse is still produced, although it is of course smaller than the original pulse. According to the specification the tolerance on the interval between $P_1$ and $P_2$ is only $\pm 0.15$ μs. Consequently, the delay line has some other tappings. One of these only 0.35 μs from the input to the delay line is used in all the decoding processes to ascertain the 'minus' tolerance. For the 'plus' tolerance limit there is a tapping at 2.45 μs associated with the 2 μs output. The signals from the outputs of the 'plus' and 'minus' tolerance tappings are passed to the suppressor inputs of the coincidence gate, and, consequently, stop decoding as soon as the tolerances are exceeded.

149

When an SLS coincidence* occurs a 35 μs wide SLS gate is produced which suppresses all the other coincidence circuits for this period by means of a coincidence suppressor.

The decoding of the interrogation modes proceeds, naturally, on the assumption that no SLS-$P_2$ pulse of sufficient amplitude has already occurred at the right time. In the instance of mode 1, the mode 1 coincidence is ascertained by means of signals at the input to and at a 3 μs tapping on the delay line provided that the mode 1 'enable' signal has been given by the control unit and has been passed on to the mode 1 gate generator. Ascertaining the 'minus' tolerance limit again takes place by means of the 0.35 μs tapping. In the remaining modes the process is similar but the tappings are at 5, 8 and 21 μs.

To limit the 'plus' tolerances the signals from the 3.6, 5.6, 8.5 and 21.7 μs tappings are connected to an OR gate so that, in the event of the tolerances being exceeded, the interrogations are rejected via the coincidence suppressor.

Decoding can also be suppressed by a trigger from the dead-time generator. At the start of the transponder reply this unit receives a pulse out of which it produces a suppressor gate approximately 100 μs in length. This thus ensures that any fresh interrogation cannot be accepted until the previous reply, including several repeated emergency signals, has been dealt with. In addition this dead-time increases the chances of an undisturbed exchange of information for the other transponders.

The decoder output signals are the mode gates and a clock gate for producing the response at the correct time.

### 3.2.5. Reference signal generator

The reference signal generator unit contains the clock pulse generator, the circuits for controlling the repetition of I/P and emergency signals, the counter circuit for the AOC overload suppressor and the output stages for the external interlock pulses.

A temperature-stabilized LC generator produces the 1.45 μs timing raster. It is externally controlled from the clock gate and oscillates strictly in phase

---

* The expression coincidence implies the simultaneous presence of the two pulses in an interrogation at the appropriate tappings on the delay line, and, after their application to an AND gate, the expression coincidence also implies the signal for the proper decoding of the associated interrogation (e.g. mode coincidence). This also applies to the $P_1$ and $P_2$ pulses which produce the SLS coincidence.

with it. After several pulse shaping and amplifier stages a rectangular timing signal is produced with an 0.45 : 1.45 mark–space ratio. This is sufficient for the timing of the coder and the control of the modulators.

### 3.2.6. The coder unit

All the other functions of the coder are grouped together in the coder unit. The influence of engineering technology on the logical layout is very obvious in this unit. Until recently, the use of transistor circuits was naturally assumed in all logic circuits. In actual fact the logical functions are designed with a combination of diode and transistor techniques. Today, such functions can be achieved with integrated semiconductor circuits. It should, however, be noted that the best use of these various techniques can only be made if their special characteristics are taken into consideration when planning the logical layout.

Thus, for example, when using integrated semiconductor circuits, the separate gating elements are very reasonably priced so that the circuit

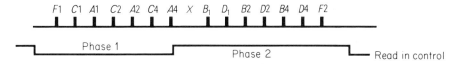

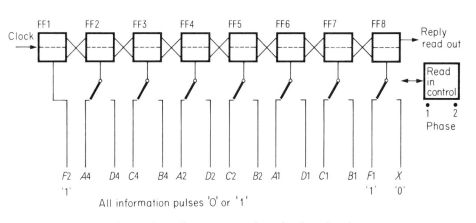

Fig. 3.13  Flow diagram for coding process passing twice through register

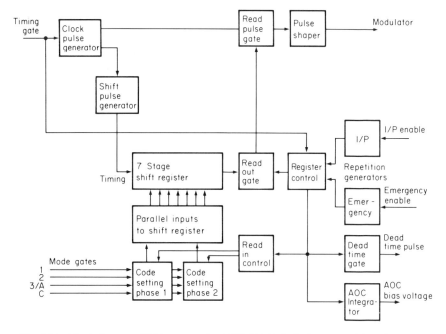

Fig. 3.14 Block diagram of IFF transponder AN/APX-46 (including parts of the reference signal generator)

development engineer can largely disregard the number of individual elements used in favour of a better arranged flow for the logical processes.

It is a different matter when designing logical gates in a discrete transistor or diode technique as, for example, in the coder. In this instance an effort has been made to produce a circuit with the minimum number of components, as is possible if the production of the reply pulse train is divided into two sections (Fig. 3.13).

Upon detecting a valid interrogation the first framing pulse F, and the information in the pulse positions A and C, associated with the reply code set in for reply to the interrogation, are 'read' in parallel, in the first 'read-in' phase, into an eight-stage shift register consisting of the stages FF1 to FF8. In the subsequent seven timing periods this information is shifted, from left to right, and is read out serially at the FF8 toggle stage. The timing of the shifting process and the 'read out' process is separated to avoid mutual interference. After the seventh timing period the register and the 'read' control are switched over to the second 'read' phase. During the shifting process in the stages FF1 to FF7, the register had already been cleared so it is now arranged in parallel with the information pulses in positions X, B, and D, and with the second framing pulse F2. The continuing timing pulses once again shift this information from left to right to supply the second part of the reply pulse trains at the output of the FF8 stage.

With these preliminary remarks the block diagram (Fig. 3.14) of the coder and reference signal generator needs no further explanation.

A register control circuit ensures that normal, I/P and emergency replies are produced in the correct sequence. Moreover, the dead-time pulse and the bias voltage for automatic overload control (AOC) are also produced in this unit.

The contents of the shift register are not yet in their final form for use as a reply. Because of variations in the transit time for each stage of the register, deviations from the exact reply raster occur. Indeed even the shapes of the pulses do not satisfy the specification. For this reason the shift register only produces control pulses which are used in the 'read pulse' gate to switch through the original clock timing pulses. By means of this artifice a correct standard series of reply pulses is produced which is free from jitter.

### 3.2.7. The transmitter unit

The transmitter unit (Fig. 3.15) contains the modulator, a self-oscillating transmitter stage (transmitter oscillator) and the associated HT generator.

The transmitter oscillator uses a disc seal triode. It is located at the end of a concentric resonator to form two tunable circuits (Fig. 3.16). Frequency is determined by the grid-cathode circuit. The grid/anode space behaves like a resonance transformer, with a secondary winding consisting of a *coupling loop* to the output.

The tuning slide for adjusting the frequency of the oscillator locks the two resonators together and also creates an electrical connection between the concentric grid/cathode cavity and the concentric grid/anode cavity.

The impedance of this connection is selected to produce positive feed-back from the output to the input circuit. An essential part of the feed-back conditions required for an oscillator is thus fulfilled. Because the anode of the transmitter valve extends out of the transmitter for cooling purposes and

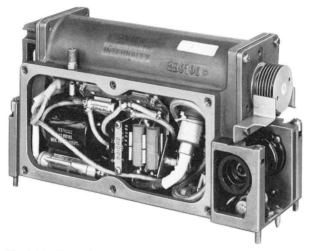

Fig. 3.15 Transmitter unit in IFF transponder AN/APX-46

is at earth potential, the HT voltage (−1500 V) is supplied to the cathode. In the quiescent state the tube is cut-off by its grid-bias voltage (−90 V relative to the cathode) and consequently does not oscillate.

To 'gate' the transmitter tube − and, therefore, to produce a pulse modulated RF oscillation − the coder signal must first be amplified in a valve circuit after which it reaches the grid of the transmitter tube via an isolating transformer. The transmitter oscillator will then oscillate as soon as, and for as long as, the positive modulating pulse is applied. With proper tuning the frequency will be 1090 MHz. The decoupled RF power reaches the antenna via the RF unit.

Particular attention must be paid to all high-voltage aircraft equipment. At high altitudes and, therefore at low atmospheric pressures the dielectric

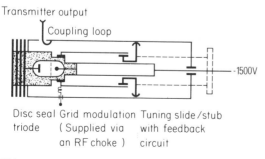

Fig. 3.16 Illustrating principle of self oscillating transmitter stage in IFF transponder AN/APX-46

strength of air falls rapidly. To avoid voltage flashovers and other discharge phenomena, the transmitter unit is designed as a casting in which all the critical circuit components, including the HT supply, are subjected to pressurized encapsulation.

At normal atmospheric pressures a valve provides the pressure balance with the environment, but, as soon the external pressure falls, the valve closes and the internal pressure remains constant.

The leakage rate through the housing including the valve, the frequency correcting elements and the sealing of the anode of the transmitter valve is so small that in the course of a flight the internal pressure largely maintains its initial value.

(Note: It is of technological interest that the transmitter is housed in a light-metal casting produced by the lost-wax process. In this process walls less than 1 mm thick can be produced, and the accuracy of the casting is so high that no subsequent machining is necessary.)

### 3.2.8. Power pack

The transponder AN/APX-46 is designed for connection to a single-phase 115 V 400 Hz mains supply. A further supply of 28 V d.c. is, however, required for the tripping relay and for the front panel illumination.

In contrast with the mains supplies provided by the electric power companies, an aircraft mains supply is subject to very heavy and sudden fluctuations in load. The mean voltage can be maintained by means of regulation within 102 V and 124 V, but switching processes with solenoids and servo drives produce very-high-voltage surges. Consequently, the power pack contains a protective circuit which reduces surges, with amplitudes of 160 V r.m.s. lasting for 700 ms and of 200 V r.m.s. lasting for 190 ms, to such an extent that the unit cannot be damaged.

The unit also contains a transformer and rectifier circuits which provide the necessary supply voltages to all the units except the transmitter. All particularly critical operating voltages are stabilized.

### 3.2.9. Test unit

When the type AN/APX-46 IFF transponders were being developed built-in test equipment (BITE) was unusual. The only provision of this type is a

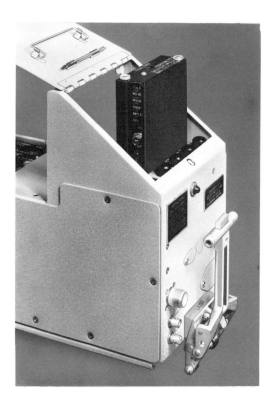

Fig. 3.17 The test unit fitted to an IFF transponder AN/APX-46

test unit for fault-finding at the most elementary stage of maintenance. This test unit can be connected to the transponder by means of a connector (Fig. 3.17). Using six indicators it is possible to test the most important functions and draw conclusions about possible sources of error:

(a) If the unit is in proper order the indicators for the mixer current and the high tension voltages must operate.
(b) By pressing a button to supply a 'noise' interrogation the transponder can be stimulated to respond (a noise signal of sufficiently-large bandwidth contains noise peaks so spaced and of such a width that they occasionally simulate a real interrogation).
(c) By means of the four other indicators it is now possible to determine whether a video signal is available (receiver test) whether a modulator trigger is available (decoder test and coder test) and whether there is any power at the transmitter output (transmitter test).

All these indications are, of course, purely qualitative.

### 3.2.10. Control unit C4083/APX

A control unit with all the necessary selector switches and adjustments is required to operate the IFF transponder. There are a number of different types whose functions are identical. The aircraft manufacturer has to select the design best suited to the cockpit lay-out. As an example, Fig. 3.18 shows the IFF control unit, type C4083/APX. The following selector switches can be recognized:

(a) The 'Master' switch with the positions 'OFF', 'STANDBY' (warming up), 'LOW' (reduced sensitivity), 'NORMAL' (normal sensitivity), and 'EMERGENCY'.
(b) The switches for interrogation modes 1, 2, 3/A, and C, by which interrogations in these modes can be accepted or rejected.
(c) The code selector switch for modes 1 and 3: mode 2 is set in on the transponder.
(d) The I/P switch.
(e) The control units for the testing and monitoring unit (see Section 3.3).

The characteristics of the IFF transponder, type AN/APX-46, are compiled in Table 7.

Fig. 3.18 IFF control unit C-4083/APX

## 3.3. Testing and Monitoring Unit T29

The secondary radar transponder is of particular importance for the safety and protection of the aircraft. In particular, when used as an IFF transponder, the decision as to whether an aircraft shall or shall not be attacked by one's own forces depends on the correct operation of the transponder. It is, moreover, extremely important that the pilot should be able to decide about the operating condition of his IFF transponder, whenever it may be necessary. By noticing a fault in good time he can break off his mission and get back safely by other means. It is for these reasons that the Testing and Monitoring Unit T29 was developed (manufactured in Europe by Siemens AG, Munich) for subsequent incorporation in all transponders which as yet had no integral (BITE) test equipment.

The testing and monitoring unit T29 (Fig. 3.19) is fitted into the antenna lead between the transponder's transmitter and receiver and the aircraft antenna. The UHF section connects the interrogation pulses produced by it into the transponder's antenna lead and extracts from it a part of the transmitter output power to investigate the most important criteria concerning the reply signals. There are, thus, two different operating conditions, namely, testing and monitoring:

(a) When testing, the transponder is interrogated in one of the four modes by an interrogation signal produced in the testing and monitoring unit T29. If the interrogation is correctly decoded the

Fig. 3.19 Testing and monitoring unit, T29

answer from the transponder can be tested for frequency, power, voltage-standing-wave-ratio, and framing-pulse spacing. The transponder can be interrogated in modes 1, 2, 3 and C, as required. The interrogation repetition frequency is approximately 450 Hz. The level of the interrogation signal is a little higher than the minimum sensitivity of the transponder. Besides determining the aforementioned four evaluation criteria it is also possible when testing to determine whether the sensitivity of the transponder's receiver is sufficient and whether it is operating at the correct receiver frequency. The T29 unit, when 'testing', will only evaluate the replies provided that the TEST button on the control unit is depressed.

The 'all correct' signal, indicated by the illumination of a green indicator lamp on the control unit, is only produced when all the above mentioned criteria are within the limits specified for the system. The 'all correct' indication on the control unit is maintained so long as at least one of the four TEST switches on the control unit is depressed, provided that the replies have been evaluated as correct.

(b) In the operation condition 'monitor', the testing and monitoring unit, T29, monitors the replies from the transponder that have been triggered by some ground station, and evaluates the transponder criteria as 'correct' or 'not correct'.

The Transponder Inflight Monitor lamp (TIM) on the control unit only indicates that the transponder is operating correctly, provided that several replies are evaluated as correct. Consequently, there is no indication of single replies produced at intervals greater than 20 ms – as may be caused, for example, by spurious voltages.

The meshing of the two test processes is shown in the block diagram (Fig. 3.20). The majority of the monitoring criteria are shown in the right-hand section. The measurement of the voltage-standing-wave-ratio, by means of a double directional coupler,* indicates whether the transmitter output power is actually being emitted by the antenna. A measurement of the framing pulse spacing is used as the criteria for a reply pulse train which correctly satisfies the specification standards. The indicator control releases the 'correct' signal if, in addition, the frequency is within the tolerance specified and if the transmitter power exceeds its minimum level.

In the operating condition 'test', on the left-hand side of the circuit, there is also a facility for coding an interrogation depending on the interrogation

---

* A double directional coupler is a line element which allows a portion of the outgoing wave, and a portion of the incoming wave, to be decoupled separately.

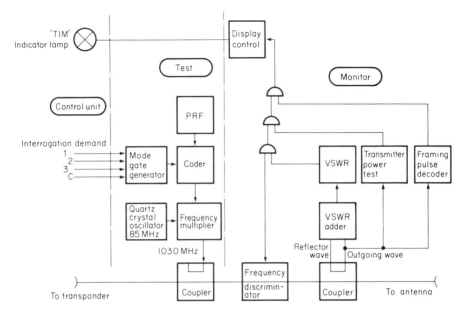

Fig. 3.20  Block diagram, testing and monitoring equipment

mode selected on the transponder. This then modulates the test transmitter which interrogates the transponder. A brief description will be given of one interesting feature of this T29 testing and monitoring unit; namely, the frequency discriminator (Fig. 3.21).

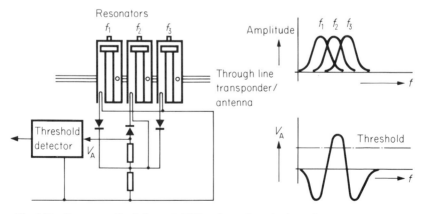

Fig. 3.21  Frequency discriminator in T29 testing and monitoring unit

Three, shortened, $\lambda/4$ resonators in the antenna lead are excited via slotted couplers. The centre resonator is tuned to the nominal frequency 1090 MHz. The other two are respectively detuned by about 8 MHz above and below the nominal frequency. Each of the three resonators has a detector output. The two resonators for 1082 and 1098 MHz produce a negative voltage at a common load resistor, but the rectifier for the centre resonator produces a positive voltage which is connected in series with the output from the other two resonators. The output voltage of this arrangement as a function of frequency is also shown in Fig. 3.21. It thus becomes clear that if the RF voltage rises, both the positive output signal and the two negative peaks will increase. This, however, implies that despite a fixed threshold it is possible to measure whether the frequency tolerance is being maintained, independently of the signal amplitude.

The exact limits for the various monitoring criteria are gathered together in Table 8.

Table 8  Limits of Monitoring Criteria

|  | Correct | Not correct |
|---|---|---|
| Transmitter frequency | W  $1090 \pm 3$ | O  $1090 \pm 5$ |
| Spacing between framing pulses $F_1$ and $F_2$ (μs) | W  $20.3 \pm 0.1$ | O  $20.3 \pm 0.3$ |
| Transmitter pulse power (dBW) | $\geqslant 400$ W | $\leqslant 200$ W |
| Voltage-standing-wave-ratio | $\leqslant 1.5 : 1$ | $\geqslant 3 : 1$ |

W = Within    O = Outside

Indirectly, this also provides information regarding the sensitivity of the receiver, the receiver frequency, as well as information on the decoding and coding by the transponder. At the correct receiver frequency (1030 MHz $\pm$ 200 KHz) the transponder must have sufficient sensitivity (at least $-69 \pm 1$ dB) to carry out the decoding correctly.

Table 9 contains the characteristics of the T29 testing and monitoring unit.

Table 9  Characteristics of Testing and Monitoring Unit, T29

| | |
|---|---|
| Power consumption | 10 W at 28 V d.c. |
| Insertion loss | ≤0.2 dB |
| Mismatching | ≤0.2 dB |
| Antenna lead impedance | 50 Ω |
| Interrogation pulse repetition frequency | 450 Hz |
| Interrogation frequency | 1030 MHz ± 200 KHz |
| Interrogation level | −53 to −80 dBm (adjustable) |
| Max. deviation within permissible temperature range | ±3 dB |
| Mode 1<br>Mode 2<br>Mode 3<br>Mode C | $3.0 \pm 0.2$ μs<br>$5.0 \pm 0.2$ μs<br>$8.0 \pm 0.2$ μs<br>$21.0 \pm 0.2$ μs |
| Pulse length | $0.8 \pm 0.1$ μs |
| Reply frequency | 1090 MHz (1080 to 1100 Hz) |
| *Other characteristics*<br>Size: Length<br>   Breadth<br>   Height | <br>10.2 cm<br>9.9 cm<br>9.3 cm |
| Altitude limit | 100 000 ft (30 000 m) |
| Temperature range<br><br>Continuous operation at sea level<br>Intermittent operation at sea level | According to MIL-E-5400 CL.2<br>55°C to +95°C depending on altitude<br>+71°C<br>+95°C |
| Humidity | 100% |
| Vibration | According to MIL-E-5400 Fig. 3<br>Curves III and IV |
| Shock | According to MIL-E-5400 |
| Control unit | e.g. C4083/APX |

## 3.4. Recent IFF Transponders

As regards the properties determined by the system, which have already been mentioned in the respective specifications, there have been no real changes in the modern IFF transponders. In some instances the tolerances for the characteristics have been reduced. Special efforts, however, have been made to increase reliability and so to guarantee the maintenance of these characteristics for longer periods of time. In this respect far more effort was required for maintenance on the older units.

The transponder briefly described below also has some novel features.

*Integral Test and Monitoring Equipment*
The principle of these so-called BITE equipments has already been explained in detail elsewhere (Section 2.3.2). The number of criteria that can be monitored is in the minimum instance similar to that provided by the T29 testing and monitoring unit, but the analysis is generally even more informative.

*Advances in Technology*
Integrated circuits (ICs) are almost exclusively used to build digital circuits and they are also being used to a considerable extent to produce analogue circuits. These components are so cheap that it is no longer necessary to develop aircraft equipment with the minimum number of components. It is now possible to develop circuits with high immunity to tolerances, and yet with higher reliability. Efforts are also made to reduce the time spent in *setting up* and maintenance. On the basis of modern techniques it is to be expected that a modern transponder will be smaller, lighter in weight, less sensitive to environmental influences, and consequently even more reliable.

### 3.4.1. IFF transponder AN/APX-90

In addition to the benefits gained from all the advances in modern technology the IFF Transponder AN/APX-90 (Hazeltine Corporation, Greenlawn, New York, U.S.A.) (Fig. 3.22) also provides a solution to the old problem of equipping an aircraft with a secondary radar antenna to permit interrogations from any direction.

As mentioned in Section 1.9.2, if a single antenna is used there is always a large sector of space in the shadow of the aircraft fuselage. By using two antennae, one on the roof and one on the belly of the aircraft, the

Fig. 3.22 IFF transponder AN/APX-90

transponder can only be switched alternatively to one or the other. In this instance the whole area is covered, but the detection probability is reduced to a half. When secondary radar was first introduced this disadvantage was less serious and could be tolerated. In the meantime, however, the density of air traffic and the speed of aircraft have increased enormously. Further, the automatic decoding equipments are designed for a flow of information, which so far as possible is continuous. Their 'intelligence' is frequently insufficient to restore contact with a previous aircraft trace after a considerable gap.

One solution to this problem, in principle by no means new, is that offered by *antenna diversity operation.*

This implies a process whereby two antenna, one on the roof and the other under the belly of the aircraft, are used (Fig. 3.23). In a 'Diversity Transponder' each of these antennae is provided with its own receiver (Fig. 3.24). A comparator at the video signal level determines through which of the two receiver channels the interrogations at the greater amplitude are arriving. Obviously at this moment, the antenna, receiving at the higher amplitude, has the better propagation conditions as regards the interrogator station, so that the reply from the transponder is therefore transmitted back via this same antenna. To achieve this an RF switch is provided. Under the control of the selector unit this switch connects the transmitter to the antenna that happens to be the more suitable.

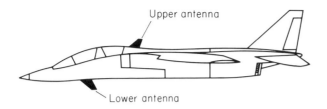

Fig. 3.23 Possible arrangement for the antennae of an SSR transponder

The IFF transponder AN/APX-90 is a real diversity transponder. Through the use of the most up-to-date technology, even including its built-in test equipment, it is smaller and lighter in weight than all the previous units of this type. The technical progress can be seen from Table 10 which compares the AN/APX-90 transponder with the AN/APX-46 model, already discussed, which was developed 13 years ago. In addition data are also given on the IFF transponder STR 700 (see Section 3.4.2).

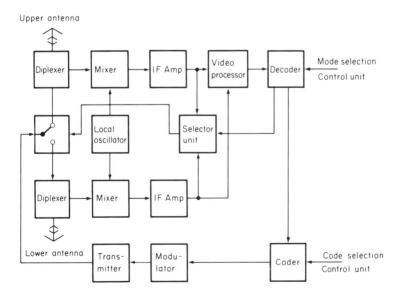

Fig. 3.24 Block diagram of SSR transponder with a two channel receiver and the equipment for 'antenna diversity operation'

Table 10  Comparison of IFF Transponders

|  | AN/APX-46 | AN/APX-90 | STR 700 |
|---|---|---|---|
| Weight (kg) | 15 without control unit | 9 without control unit | 10 with control unit |
| Dimensions (cm) B × H × L | 18 × 33 × 61 | 14 × 13 × 31 | 14 × 19 × 32<br>15 × 14 × 16<br>(Control unit) |
| Volume (l) | 36 | 5.7 | 8.5<br>3.5 |
| Environment as per MIL-E-5400 | Class 2 | Class 3 | Class 2+ |
| Temperature range<br>Intermittent<br>Vibration | −55°C   +71°C<br>+95°C | −55°C   +95°C<br>+125°C<br>Curve IV (without shock absorbers) (10 g) | −55°C   +95°C<br><br>Curve IV (with shock absorbers) (10 g) |
| Receiver<br>Diversity<br>Dynamic range<br>Sensitivity | <br>No<br>50 dB<br>−75 dBm | <br>Yes<br>60 dB Log IF Amp.<br>−77 dBm | <br>Yes<br>60 dB Log IF Amp.<br>−75 dBm |
| Transmitter<br>Peak pulse power<br>Coder and decoder | <br>500 W<br>As modified to suit AIMS/64 specifications | <br>800 W<br>As per AIMS | <br>500 W<br>As per AIMS |
| BITE (test equipment) | Only with external unit T29 | Yes. Approximately twice as many functions as T29 | Yes. Approximately twice as many functions as T29 |
| Power supplies | 102 ... 124 V/400 Hz 90 VA and 28 V d.c./5 W | 17 ... 30 V d.c.   80 W | 17 ... 32 V d.c.   80 W |

### 3.4.2. IFF transponder STR 700

The IFF Transponder STR 700, which has been developed by Siemens AG, Munich, is also a diversity transponder. It has been designed to suit the special operating conditions in Europe. Although the electrical characteristics are largely the same as those of other transponders, e.g. AN/APX-90, the stress laid on environmental requirements has only been such as to satisfy the requirements of aircraft currently available or such as will shortly be planned. These requirements are roughly intermediate, between those of Classes 2 and 3 of the military specification for aircraft equipments. This limitation has facilitated remarkable reductions in costs.

In the instance of the STR 700, efforts have also been made to achieve convertibility so as to be able to preserve the advantages of mass production despite use in very different types of aircraft.

Fig. 3.25   IFF transponder STR 700

For one of the possible arrangements the unit was divided into two sections such that the actual transponder only contained the transmitter, the two receivers with the evaluating unit, and the power pack (Fig. 3.25). The coder and decoder are incorporated in the control unit. This solution avoids many of the interconnection leads from control unit to transponder. Consequently, on an aircraft of medium size a saving in weight of cabling of as much as 5 kg can be achieved. If the weight of the rest of the transponder equipment is estimated at 10 kg, such a saving in weight is no trivial matter. The other characteristics which distinguish the transponder STR 700 from older units have already been mentioned in the table of comparisons for the AN/APX transponders.

Whilst, in the past, analogue delay lines have been used almost exclusively for series parallel conversion during the decoding process, a digital shift-register is used in the STR 700 decoder. The interlace to the remaining logic circuits by this no longer poses a problem. Indeed in the future this method will bring about further improvements through large scale integration.

(Note: For physical reasons the size of an analogue delay line can hardly be reduced any further. In the future it will thus be so clumsy as to represent a considerable anomaly when compared with the technology of the rest of the equipment.)

The problem of digital decoding is concerned with the fact that an interrogation can arrive at any time, whilst a shift-register can only work to a fixed timing.

The obvious solution is to start a clock pulse generator with the correct phasing on receiving the first interrogation pulse, $P_1$, just as was done in the reference signal generator of the AN/APX-46. If one knew that an incoming pulse really was the $P_1$ pulse of an interrogation, and that a further $P_3$ pulse would follow it, then this solution would actually be possible. If, however, the incoming pulse was only a single spurious pulse – and there are many such – then the clock pulse generator would be started uselessly and no further interrogations could be accepted for a period of 25 µs.

There is another possibility; namely, that of using a continuously operating shift pulse mechanism. Of course, since pulses can enter in any phase relationship, there will be a timing error in the digitalization. The block diagram (Fig. 3.26) shows how this problem can be avoided.

The video signal is first examined in a pulse analyser to determine whether it is in accordance with standard interrogation pulses as regards its width and other properties. A number of the spurious pulses are, in this way, eliminated at the start.

When it has been established that the first pulse to arrive seems to be an interrogator pulse, it can be converted for further processing in a pulse

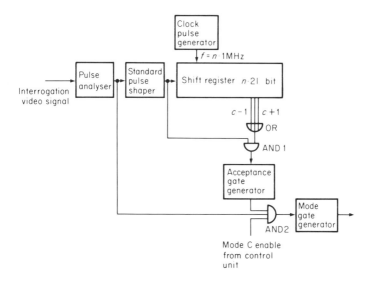

Fig. 3.26 Block diagram showing principle of digital decoding, using Mode C as an example

shaper, into a normal pulse, with the width of one bit for the shift-register, so that it has the leading edge and time conditions of a standard pulse. This standard pulse is then fed to the input of a shift-register, through which it is shifted.

Depending on the interrogation mode the intervals to be decoded are, 2, 3, 5, 8 and 21 µs. The L.C.M. of this series of numbers is 1 µs. At a shift timing of 1 MHz, corresponding to a timing of 1 µs, the error in digitalization would, of course, be too large. Consequently, a timing frequency of $n \times 1$ MHz is chosen.

The capacity of the shift-register must accordingly be increased by the factor $n$. The factor $n$ cannot be made just as large as one pleases. The choice involves the error in digitalization, acceptance and rejection tolerances, and costs.

After a given period of time there will appear at the output C, in the example of the shift pulse chosen, a pulse with the pulse interval for mode C, i.e. 21 µs. On leaving the shift-register this pulse is applied to an OR gate. Because the interval between $P_1$ and $P_3$, however, will seldom be exactly 21 µs and since deviations of up to $-0.2$ µs are permitted by the system, and with respect to the error in digitalization of $1/n$ µs maximum, the pulse, as it passes through the register, is also picked off from the adjacent storage units, $C-1$ and $C+1$, for supply to the OR gate. Its output therefore produces a gate covering all the possible tolerances in pulse position. This gating pulse is compared for coincidence in an AND gate with the standard $P_3$ interrogation pulse which will now be present at the input to the shift-register. If the choice of shift timing has been correct this can only happen where the spacing between $P_1$ and $P_3$ is within the permissible tolerance of $\pm 0.2$ µs. It certainly will not happen if the time interval, in accordance with the rejection level, deviates more than 0.8 µs from the standard value.

If there is an acceptance criterion, then this is a means of providing an acceptance gate which covers the width of the original pulse, the permissible tolerances and the possible errors in digitalization. Therefore, if it has received the 'enable' for mode C from the control unit, this acceptance gate pulse then switches the *original* $P_3$ pulse in its correct position in time, via a second AND gate, through to the mode gate generator. In actual practice, of course, the acceptance gate pulse would arrive too late for the original pulse owing to the differences in travel times. This can, however, easily be compensated, by making the shift-register a little shorter and by delaying the original $P_3$ pulse for a short time. In this way it is possible to produce a mode gate by digital decoding which is absolutely 'jitter free'.

Table 11  Compilation of the most important modern transponders

| Manufacturer | ARC | BENDIX | BENDIX | BENDIX |
|---|---|---|---|---|
| Identification mark | RT-506A | TRA 61 | TPR-600 | TPR-610 |
| *Manufactured* in accordance with specification | TSO, C74 Cat. BA  AAAE Class 1 | TSO, C74a Cat. DAXXXX Class 1 | TSO, C74 Cat. BAAAAAX Class 1 | TSO, C74 Cat. DAAAAX Class 2 |
| *Transponder* Weight (kg) Dimensions B × H × L (cm) Volume (l) | 1.9  9.9 × 10.1 × 13.1 3.1 | 6.8  12.4 × 19.4 × 32 7.7 | 2.8  9.8 × 12.6 × 34.6 4.3 | 1.72 (High model)  8.1 × 6.6 × 30.5 1.63 |
| *Control unit* Weight (kg) Dimensions B × H × L (cm) Volume (l) | 0.272  9.1 × 6.4 × 8.2 0.5 | 1  14.4 × 6.4 × 10 0.9 | 0.45  6.2 × 7.3 × 12.2 0.55 | 1.81 (Flat model)  16.5 × 3.8 × 30.5 1.91 |
| *Environmental influences* | | | | |
| Celsius temp. (intermittent) Altitude (m) | −46 to +55 (+71) 9150 | −46 to +55 (+71) 9150 | −46 to +55 (+71) 13700 or 9150 Type A    Type B | −15 to +55 (+60) 4570 |
| *Receiver/type* Oscillator stability Sensitivity  normal  low Dynamic range Selectivity SLS | Superheterodyne receiver   −76 dBm   <3 dB at ±2 MHz >60 dB ±25 MHz 3 pulse process | Superheterodyne receiver Quartz crystal controlled  −74 dBm between −69 and −34 dBm 50 dB 6 MHz < 3 dB 3 pulse process 2 pulse, if required | Superheterodyne receiver   −72 dBm    3 pulse process | Superheterodyne rece Quartz crystal contro  −74 dBm at 90% −62 dBm  50 dB log. 3 pulse process |
| *Decoder/type* | Monostable multi-vibrators | | | Delay line |
| Modes | 3/A C | A, B, C, D | A, C | A (B) C |
| Coder type | 9 stage shift-register | | | Delay line |
| Mode A | 4096 + SPI | 4096 + SPI | 4096 + SPI | A, B: 4096 + SPI |
| Mode C | MoA-Gilham | MoA-Gilham | MoA-Gilham | MoA-Gilham |
| *Transmitter*  Frequency/stability Transmitter power (W) Reply rate | Self excited output stage 1090 MHz 500 1200 replies s⁻¹ | 1090 ± 2.5 MHz 500 | Oscillator and amplifier 1090 MHz 500 ± 3 dB AOC 1200 replies s⁻¹ | Self excited output stage 1090 ± 3 MHz 200 ± 3 dB AOC 1200 replies s⁻¹ |
| *Power supplies* Voltage/Current (V/A)  Power consumption (W) | 14/3 Type 0014 28/1.5 Type 0028 42 | 115 V, 300–1000 Hz 55 VA Or 28 V/2 A, 55 W | 14/2.6 or 27/5/1.3 36.4 | 14/0.8 Type 01 28/0.4 Type 02 11.2 |
| *Test equipment* | Self-Test-Button (Control unit) 'Go-no-go' indicator | Monitor and self test | 'Go-no-go' indicator | Self-Test-Button 'Go-no-go' indicator if required |
| *Accessories* | Control unit: C-506 A  Antenna: A-105 B Shock mount | Control unit CNA 61 A  Antenna: ANA 29/61 A | Control unit: CN-602 A : CN-602 B : CN-602 C Antenna: AT-604 A Shock mount MT-603 A | Antenna: AT-914 A |
| *Special features* | 1 tube (transmitter)  Thick film circuits | | 2 tubes and transistors With control unit CN-602 C 2 transponders can be controlled | 1 tube (transmitter)  Control unit and transponder are a single unit |

| OLLINS | COSSOR | GENAVE | KING | KING |
|---|---|---|---|---|
| 21 A-6 | SSR-2100 | Beta 4096 | KXP-750 | KT 75 |
| SO, C74a<br>RINC 572 | ICAO<br>ARINC 532 D<br>Section 3, 8, App. 10 | TSO, C74a | TSO, C74<br>Cat. B/DAAAAAX<br>Class 1 | TSO, C74a<br>Cat. B/DCABBX<br>Class 1 |
| | Transmitter/Receiver<br>1.8 | 2 | 2.7 | 3.17 |
| 19.4 × 32 | 5.7 × 12.4 × 20.3<br>1.43 | 16.5 × 5 × 30.5<br>2.4 | 3.3 × 13.4 × 30.5<br>3.4 | 15.3 × 6.4 × 28<br>2.7 |
| | with Coder/Decoder<br>0.77 | | 0.36 | |
| | 14.6 × 5.7 × 10.2<br>0.85 | | 6.2 × 6.4 × 6.4<br>0.25 | |
| | B.C.A.R. Cat I<br>BSG 2G 100 Part II<br>−26 to +55<br>12 200 | | −46 to +55<br>(+71)<br>9150/4570 | −46 to +55<br>(+71)<br>9150/4570 |
| perheterodyne receiver<br>artz crystal controlled | Superheterodyne receiver<br>Quartz crystal controlled | Superheterodyne receiver<br>Quartz crystal controlled | | Superheterodyne receiver<br>Quartz crystal controlled |
| 6 dB | −77 dBm at 90% | −72 dBm | | |
| log characteristic<br>dB at ± 4 MHz<br>0 dB at ± 25 MHz<br>ulse process | 50 dB log.<br>>6 MHz at −3 dB<br><50 MHz at −60 dB<br>3 pulse process | 3 pulse process | | 3 pulse process |
| ital shift-register | Digital shift-register | | | Monostable<br>multi-vibrators |
| 3, C, D | A, B, C (D) | A, C | A, C | A, C |
| ital shift-register<br>artz controlled timing<br>6 + SPI | Digital shift-register | 4096 + SPI | 4096 + SPI | IC shift-register<br>4096 + SPI |
| A-Gilham | MoA-Gilham | MoA-Gilham | MoA-Gilham | MoA-Gilham |
| excited output<br>e<br>0 MHz<br>Duty cycle | Self excited output<br>stage<br>1090 ± 3 MHz<br>>500<br>18 000 pulses s$^{-1}$ | 1090 ± 3 MHz<br>200 ± 3 dB | | 1090 MHz<br>300 |
| V. 400 Hz<br>/A | 28/1.75<br>49 | 14/2.4 max<br>34 | 28/1<br>28 | 14/2.4 or switch to<br>28/1.2<br>33.6 |
| nsive automatic<br>equipment | Self-Test-Button<br>'Go-no-go' indicator | Automatic test equipment | Self-Test-Button | Reply-light |
| trol unit | Control unit | | Control unit: KFS 570 | |
| enna | | Antenna<br>lambda 1000 | Antenna<br>Shock mount | Antenna: KA 28 |
| | 1 tube (transmitter-<br>ceramic triode)<br>Transponder divided<br>into 2 units | | 1 tube (transmitter) | 1 tube (transmitter)<br>integrated circuits<br>and transistors.<br>Control unit and<br>transponders form one<br>unit |

Table 11—cont.

| Manufacturer | NARCO | NARCO | RCA | WILCOX |
|---|---|---|---|---|
| Identification mark | UAT 1 | AT6-A | AVQ-65 | Model 1014 A |
| *Manufactured* in accordance with specification | TSO, C74 Cat. BAEAAAX Class 1 | Gen. FCC and FAA Specifications | ARINC 532D TSO, C74a Cat. BAAAAAX Class 1 | TSO, C74a Cat. AAAAAX Class 1 |
| *Transponder* Weight (kg) Dimensions B × H × L (cm) Volume (l) | 3.44<br>9.5 × 15.2 × 34.2<br>5 | 2.21<br>7.6 × 12.7 × 29.6<br>2.86 | 4.7<br>9.4 × 19.4 × 38.5<br>7 | 2.55<br>6.7 × 12.7 × 29.9<br>2.6 |
| *Control unit* Weight (kg) Dimensions B × H × L (cm) Volume (l) | 0.23<br>6 × 7.3 × 12.7<br>0.55 | 0.9<br>16 × 3.8 × 19<br>1.15 | 0.45<br>5.7 × 8.7 × 7.3<br>0.36 | 0.254<br>6.7 × 7.3 × 10.8<br>0.55 |
| *Environmental influences* Celsius temp. (intermittent) Altitude (m) | −46 to +55 (+71)<br>9150 | 9150 | −46 to +55 (+71)<br>13700 | −55 to +71<br>13700 |
| *Receiver/type* Oscillator stability Sensitivity   normal   low Dynamic range Selectivity<br><br>SLS | Superheterodyne receiver Quartz crystal controlled<br><br>−75 dBm<br>−63 dBm<br>50 dB log.<br><3 dB at ±2 MHz<br>>60 dB at ±25 MHz<br>Available | Superheterodyne receiver Quartz crystal controlled<br><br>−72 dBm<br><br>50 dB log.<br><3 dB at ±2 MHz<br>>60 dB at ±25 MHz<br>3 pulse process | Superheterodyne receiver Quartz crystal controlled<br><br>−74 dBm<br>−62 dBm<br><br><3 dB at ±3.5 MHz<br>>60 dB at ±25 MHz<br>3 pulse process<br>(2 and 3 if required) | Superheterodyne recei− Quartz crystal control−<br><br>−74 dBm at 90%<br>−62 dBm<br><br><3dB at 3 MHz<br>>60 dB at 25 MHz<br>3 pulse process |
| *Decoder/type* | Delay lines with buffer stages | Mode A delay line Mode C monostable multi-vibrators | | |
| *Modes* | A, C | A, C | A, B, C, D | A, C |
| *Coder type* Mode A Mode C | Binary counter with diode matrix<br>4096 + SPI<br>MoA-Gilham | Dynamic shift-register<br><br>4096 + SPI<br>MoA-Gilham | <br>A, B, D: 4096 + SPI<br>MoA-Gilham | Delay line<br><br>4096 + SPI<br>MoA-Gilham |
| *Transmitter*<br>Frequency/stability Transmitter power (W) Reply rate | Quartz crystal controlled<br>1090 MHz<br>500 | Self excited output stage<br>1090 MHz<br>150 | Self excited output stage<br>1090 ± 2.5 MHz<br>500 | Self excited output stage<br>1090 ± 3 MHz<br>500 ± 3 dB |
| *Power supplies* Voltage/Current (V/A)<br><br>Power consumption (W) | 13.75/3.5 or switch to 27.5/1.75<br>48 | 11 ... 32 V without switch at 13.75/2<br>at 27.5/1.4<br>38 | 27.5/1<br>27.5 | 12 ... 30 V at 13.75/2<br>at 27.5/1.5<br>max 48 |
| *Test equipment* | Test position on main switch<br>'Go-no-go' indicator | Test position on main switch<br>'Go-no-go' indicator | Test position on main switch<br>'Go-no-go' indicator | Test-monitor switch<br>'Go-no-go' indicator |
| *Accessories* | Control unit: UTC 1<br><br>Antenna: UDA 3<br>Shock mount | <br><br>Antenna: UDA 2<br>Shock mount | Control unit<br><br>Antenna<br>Shock mount | Control: 97707<br><br>Antenna<br>Shock mount |
| *Special features* | Quartz crystal stabilized receiver oscillator and transmitter | Transponder divided into two units (control unit and transponder). Mains supply with switching regulator | 1 tube (transmitter) Different types for connection to ARINC control unit and ARINC equipment mounting | 1 tube (transmitter)<br><br>Stripline filter Switched regulator |

## 3.5. SSR Transponder for Air Traffic Control

As we have already mentioned, the possible applications for SSR transponders in civil aviation are numerous. At a first glance one would also be inclined to make a sub-division in this field, into transponders for the large aircraft belonging to commercial airlines and transponders for the small general aviation aircraft.

Further consideration, however, shows that such a proposal is not valid, and that transponders can best be grouped according to a price category. Such a classification is sensible since one can assume that the price for any unit with standard characteristics gives a good indication of its quality, its reliability, and the amount of technical effort devoted to it. Since it is very difficult to find a common price basis, any statement on price categories can only be approximate. Accordingly the most important modern SSR transponders are quoted, in Table 11, in alphabetical order.

This table makes it clear that *all* the transponders satisfy the minimum requirements, but that, as a result of the various different categories of usage, the available supplies can be sub-divided into a number of types.

A further method of classification is offered by the manufacturers' efforts to achieve the best compromise between technical and economic requirements in relation to all the other aircraft equipment. This affects the method of construction, the types of components used, and, of course, the price.

To illustrate these possible different methods of classification in greater detail, the following sections will give illustrations of seven SSR transponders in different price categories. Each type will therefore be used to illustrate a special design characteristic in greater detail. In this way it will be possible to obtain a reliable survey of the various techniques adopted without having to discuss repeatedly those parts of the circuits that are already well known.

### 3.5.1. ATC transponder AVQ-65

The AVQ-65 transponder (a product of RCA, the Radio Corporation of America, Aviation Equipment Department, Los Angeles, U.S.A.) which is in the high-price category (Fig. 3.27), is principally for use in commercial aviation as well as for average-sized private aircraft. One arrangement has been specially designed to satisfy the ARINC specification 532D. The actual transponder is housed in a *standard* ARINC '$\frac{3}{8}$ ATR short' rack.

Fig. 3.27 ATC transponder AVQ-65 made by the Radio Corporation of America

Depending upon the installation location, two types of control unit are available, which only contain the switching and setting components and the indicator lamps. A comparison with the control unit for a military IFF transponder, described in Section 3.3.10, makes it clear that the number of control items required for civil air traffic control is considerably smaller. Use is usually made of only one recognition code in one of the interrogation modes A, B or D. Altitude information can also be transmitted. The testing and monitoring equipment only operates in the mode being used.

A toggle switch in the left, upper corner of the control unit, which has the inscriptions No. 1 and No. 2, is used to make the connections to the two transponders in aircraft equipped with a dual installation to increase operational reliability. If the built-in test equipment indicates to the pilot that the transponder actually in use is not functioning correctly, he can transfer to the reserve transponder, simply by operating the switch.

### 3.5.2. ATC transponder SSR 2100

The transponder SSR 2100 (made by A. C. Cossor Ltd, Harlow, Essex, England) also belongs to the high-price category. It also is mainly used in large and average-sized aircraft. Thus, for example, the following types of aircraft are equipped with a transponder SSR 2100: BAC 111, Series 500, Trident 3B, Concorde, Viscount Jetstream and Queen Air.

Fig. 3.28  ATC transponder SSR 2100 made by Cossor Electronics Ltd

The essential characteristic of the SSR 2100 is that the coder and decoder are housed in the control unit. The fact that this is possible is principally because the digital decoding is achieved by using integrated circuits in a flatpack package. In comparison with the circuit described in Section 3.4.2 the decoder has another circuit modification to improve reliability.

In a digital register the remainder of the information in the register is generally lost if one single stage should fail. As the length of the register increases, therefore, its reliability rapidly decreases.

This problem is avoided in the SSR 2100 transponder by means of a 'redundancy' arrangement whereby the information is stored in two parallel registers, A and B (Fig. 3.29). The capacity of each register is exactly half that of a single series register.

When using these two shorter registers the full delay-time has to be allowed for by halving the frequency of the clock pulses. To prevent any increase in the digitalization error the registers are alternately driven with a phase shift of 180°, and the input information is alternately recorded in the A and B registers. The information from the two registers is recombined in a 'readout' circuit after which it can be processed in the manner already described.

If one bit in one of the two registers should now fail, the other register will still operate and the proper delay required for decoding will still be maintained. It is only the digitalization error which is increased. The

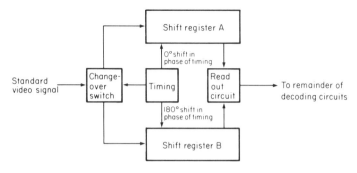

Fig. 3.29 Increasing reliability of the shift-register in the decoder by connecting in parallel two separate registers, A and B, with half the capacity and shifting the phase of the timing pulses by 180°

acceptance tolerances which are critical for the evaluation of an interrogation are largely preserved. The rejection tolerance for interrogations which do not meet the system standards are slightly increased but this is not as serious a matter as the complete failure of the unit.

### 3.5.3. ATC transponder 506A

The ATC transponder 506A (Fig. 3.30) (made by the Aircraft Radio Corporation, Boonton, New York, U.S.A.) belongs to the medium-priced

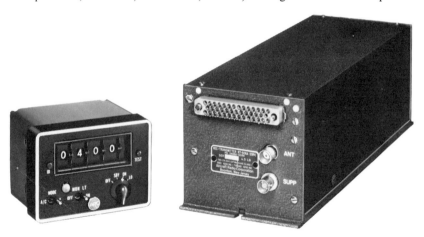

Fig. 3.30 ATC transponder 506-A made by the Aircraft Radio Corporation

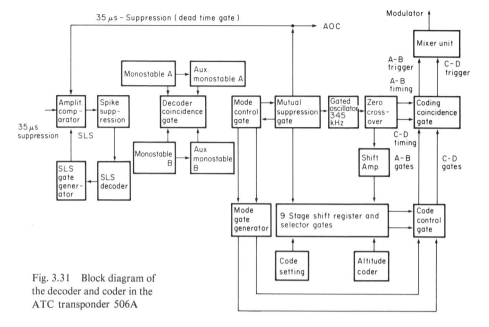

Fig. 3.31 Block diagram of the decoder and coder in the ATC transponder 506A

category. During the development of units in this class particular consideration had to be paid to the problem of reducing the cost of the circuitry by skilful design ideas. Thus, the transponder 506A has an original solution for the decoder and coder section, the most striking feature being the use of monostable multi-vibrator in the decoder logic.

The block diagram (Fig. 3.31) shows how the signal passes, as usual, through an amplitude comparator to suppress any echoes and through a discriminator for spike suppression. After this point one part of the circuit branches off to a side-lobe suppression circuit.

This circuit consists, basically, of a monostable multi-vibrator with a delay of 1.2 μs which corresponds to the minimum acceptance tolerance for SLS pulses. At the output of the monostable multi-vibrator stage there is an $RC$ circuit to control the threshold value circuit. If a second pulse appears after $P_1$ (within 1.2 μs and 2.8 μs, the discharge limit of the $RC$ circuit), then this pulse will be switched through to the SLS gate switch, also a monostable multi-vibrator, which triggers the SLS dead time (35 μs). The acceptance of $P_3$ is then prevented by the usual method.

Simultaneously, $P_1$ also reaches the actual decoding circuit, that is to the inputs to the monostable multi-vibrator stages for modes 3/A and C. As the

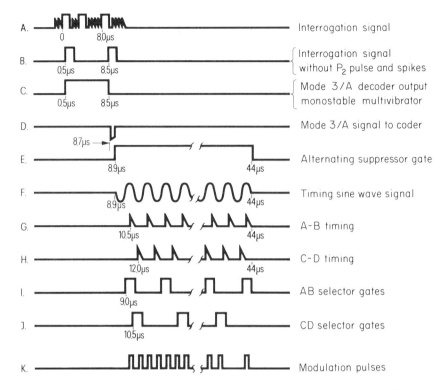

Fig. 3.32 Pulse timing diagram for the decoder and coder in the ATC transponder 506A

pulse-timing diagram (Fig. 3.32) shows, these monostable multi-vibrators are set in operation by $P_1$ and cease to operate after a specially-adjusted period: after 8 μs, in the instance of the mode 3/A detector, and after 21 μs in the instance of the mode C detector. The trailing edges of the pulses are differentiated and amplified. If the $P_3$ interrogation pulse now follows at the correct time, then the coincidence of the two pulses is an indication of the presence of an interrogation. The mode gates are therefore produced in the mode gate generators and output via the mode control gate. The mode control gates, however, also trigger a mutual suppressor gate, the function of which is, on the one hand, to initiate a reception dead time throughout the remainder of the decoding process and, on the other hand, to gate a basic timing generator with a frequency of approximately 345 kHz (corresponding to a pulse interval of 2.9 μs). This basic timing generator is phase-locked to the mutual suppressor gate and supplies a sinusoidal oscillation to a zero transition detector, which supplies an output pulse for each zero transition. The output pulses occurring during the rising transition are used for the A and B pulses whilst those occurring during the falling transition are used for the C and D pulses.

By thus dividing the coding process into two phases it is possible to carry out the coding process in a single nine-stage shift-register. The individual information pulses are kept separate in coding mode control gates and in coding coincidence gates. The separately produced reply signals in the A—B and C—D pulse groups are then supplied to an OR circuit whence they go to the modulator.

(Note: When decoding with monostable multi-vibrators a further point must be maintained. Where there are violent fluctuations in the environmental conditions the delay time of such monostable multi-vibrators can fluctuate very considerably from the nominal value. The transponder 506A consequently uses integrated components, manufactured by the *thick film technique,* using *carefully-selected discrete components.* In this way the required decoder tolerances can be maintained.)

When using monostable multi-vibrators, there is the possibility that one multi-vibrator stage can be triggered by a spurious pulse. This stage is then blocked for the remainder of the adjusted delay time. So that a desired pulse, appearing later, should be effective, a second identical monostable multi-vibrator is supplied with each channel and is always connected to the input, if the first multi-vibrator is actuated. To be absolutely certain that no desired interrogation is lost, further reserve channels can be connected in parallel. However, even with two channels, the decoder losses are (in practice) sufficiently few.

### 3.5.4. ATC transponder UAT-1

The transponder UAT-1 (Fig. 3.33) (made by Narco Avionics, a division of Narco Scientific Industries, Fort Washington, Pa, U.S.A.) belongs to the 'Custom line' family of instruments. A whole series of aircraft instruments similar in layout using the same design principles, belonging to this family, can be supplied. The transponder is in the medium-price category. It is housed in the standard '$\frac{3}{8}$ ATR short' housing.

The transmitter design is the most remarkable feature of the transponder UAT-1. The $\pm 3$ MHz tolerance in the transmitter frequency of an SSR transponder is intentionally wide. It thus becomes possible, in principle, to

Fig. 3.33   ATC transponder UAT-1 made by Narco Avionics

use self-oscillating power output stages. This solution is therefore used in the majority of transponders. However, it involves certain problems. Due to ageing of the valves and faulty antenna matching, the frequency tolerance is often exceeded. Consequently, a considerable amount of time has yet to be devoted to the development and manufacture of transmitter stages which will be able to operate within the limits of the given frequency tolerance. Particular attention also has to be given to the fact that in actual operation checks to find out whether the transmitter frequency is within the permissible limits have to be made rather too frequently. This can be a source of difficulty to general-aviation aircraft at secluded home airports.

To avoid such difficulties from the start, the decision was made to use a quartz-crystal-stabilized transmitter in the UAT-1 transponder. The block diagram (Fig. 3.34) illustrates the basic idea.

A quartz-crystal oscillator supplies a reference frequency of approximately 91 MHz. Part of the output power is supplied to the receiver mixer via two, transistorized, frequency-doubler and a diode tripler. *En route* to the transmitter, the 91 MHz signal is first sampled in a transistor stage and is then applied to a frequency-doubling circuit from which the output signal (at 182 MHz) is supplied to a frequency tripler.

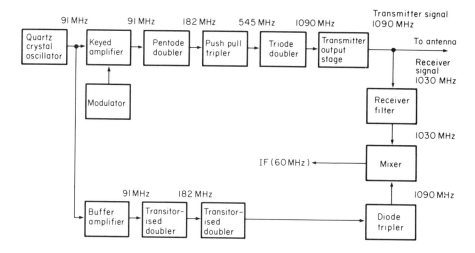

Fig. 3.34 Block diagram illustrating the processing of the transmitter and receiver frequencies in the ATC transponder UAT-1

The following stages, frequency doubler and transmitter output stage, consist of coaxial circuits, fitted with disc-seal triodes, the assembly of which is mechanically identical.

The receiver in the UAT-1 transponder is transistorized. The IF signal from the mixer is first fed to an IF amplifier which also controls the gain control for the automatic overload control circuit. It is then connected to a six-stage transistorized IF amplifier, five stages of which form a logarithmic amplifier operating on the successive-detection principle (explained in Section 2.4.5).

The RF part of the UAT-1 transponder is a very carefully designed and expensive item. In seeking to achieve a cost-effective solution a very unusual design has been chosen for the coder and decoder section.

This function of this decoder returns to that of an analogue delay unit. This, however, is not the compact delay-line, with tappings, with which we have become familiar in other transponders. Actually, it is composed of several single, low-pass-filter units with delays of 2, 3, 3, 3, 3, 3, 3, and 1 μs. Each of the stages is decoupled by an amplifier stage to regenerate the video signal.

The method of coding also differs from methods previously discussed. The main details of this method are illustrated in a block diagram (Fig. 3.35).

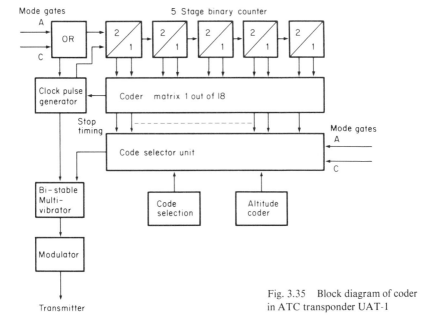

Fig. 3.35 Block diagram of coder in ATC transponder UAT-1

A mode gate for mode A or mode C is the signal that an interrogation has been identified as correct. The output of a clock pulse generator, which is started from the output of an OR gate, triggers a five-stage binary counter. The output signals from the binary counter are not yet, however, immediately suitable for the production of the pulsed reply telegram which, to include the framing pulses and the SPI pulse, requires 18 serially-arranged uniform coding stages.

For this purpose the binary number must first be re-encoded, in a coding matrix, into a serial signal suitable for coding, i.e. into a '1 from 18' code. This '1 from 18' signal can then be connected to further groups of logical units set, on the one hand, by the control unit and, on the other hand, set by the code-selector units under the control of the mode gates which will then form the reply pulse train.

The nineteenth step of the counter, subsequent upon the pulse train, stops the clock-pulse generator and resets the counter.

The clock-pulse generator also illustrates (Fig. 3.36) a method of solving the problem of an externally-controlled generator which has not as yet been mentioned.

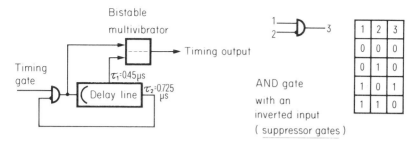

Fig. 3.36 Block diagram of clock-pulse generator using a delay line

If a timing-gate pulse at a potential 1 is applied to the suppressor gate, then a 1 condition will also be present at the output of the as yet unsuppressed logical gate. This signal passes through the delay line and, after an interval of 0.725 μs, appears at the suppressor input of the logical gate. The output from the logical gate is thus reduced to the 0 potential. This 0 potential, which will terminate the suppressor pulse, also passes through the delay line, and, after 0.725 μs, once again removes the suppression at the input of the logical gate. Since the timing gate pulse is always at 1 the above process repeats itself once again. As an output signal a rectangular signal with a 1 : 1 mark–space ratio, and a timing of $2 \times 0.725 = 1.45$ μs can be taken from the suppressor gate. To obtain a standardized output signal, the leading edge of this oscillary signal is used to set a bistable multi-vibrator which is reset by a signal provided from the 0.45 μs tapping on the delay line. This thus produces the basic timing raster with pulses 0.45 μs wide at intervals of 1.45 μs.

### 3.5.5. ATC transponder AT 6-A

The 'Flightguard' transponder AT 6-A (Fig. 3.37) (made by the Narco Avionics Division of Narco Scientific Industries, Fort Washington, Pa, U.S.A.) belongs to the lower-priced category. The equipment is divided into a transmitter-receiver unit and a control unit, which contains the coder and decoder. These two sections are connected, via a multiplexing device, by a common coaxial cable for interrogations, replies, and remote control. This equipment is much more simply constructed than the UAT-1 transponder. The transmitter, for example, is a grid-modulated self-oscillating power stage. The interrogations are received by a superheterodyne receiver with a quartz-crystal-stabilized local oscillator (Fig. 3.38). The 60 MHz

Fig. 3.37 ATC transponder AT 6-A made by Narco Avionics

intermediate frequency is amplified in a three-stage linear (transistor) amplifier. To ensure that the input signal will have a dynamic range sufficient for decoding, logarithmation of the interrogator signal takes place after detection – that is, at the video level – by means of a diode circuit (see Section 2.4.5).

This logarithmation circuit causes a compression of approximately 40 dB, and, with powerful interrogator signals, saturation of the second and third IF amplifier stages produces another 10 dB compression.

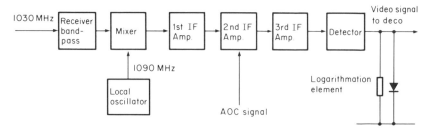

Fig. 3.38 Block diagram of receiver in ATC transponder AT 6-A

The decoding of the interrogations is achieved by combining two processes which have already been dealt with in other transponders.

The SLS criterion and interrogations in mode 3/A are decoded by using low-pass filters as delay elements in one 2 μs section and two sections each of 3 μs length separated by buffer amplifiers.

To decode mode C interrogations a monostable multi-vibrator is used to cover the periods from 8 μs to 21 μs. A reserve stage is also connected in parallel with this stage, in case the first multi-vibrator is triggered by a spurious pulse. The advantage of this combination of delay line and monostable multi-vibrator is that the SLS function and the decoding of the identification mode 3/A are to a large extent insensitive to interference whilst the costs of the long delay time are reduced for the decoding of the less-important mode C.

A special shift-register is used for coding on the AT 6-A transponder. The usual type of shift-register can also be termed 'static' registers. As long as the operating voltage is applied, they store the information 'written into' them. This information can be shifted in accordance with any required timing, it can be renewed or erased. If the clock pulses are switched-off, the latest available information state is preserved. To create such a shift-register involved considerable expense on special circuit techniques.

However, the requirements for coding a secondary radar reply are much less stringent. The shift timing has a specified value of 1.45 μs. Once a coding process has been started, it continues uniformly until its conclusion when the register must be reset. These conditions can be fulfilled by a circuit known as a 'dynamic' shift-register, which only operates within a specified range of shift rates. If the register is operated at a timing frequency lower than the specified minimum, or if gaps occur in the timing pulses, then the information recorded in the register is erased. This disadvantage is compensated by the considerable reduction in the cost of the components as compared with the 'static register'.

This is illustrated by a section of the coder circuit diagram (Fig. 3.39). The mode gates are used, as usual, to control a clock-pulse generator which indirectly controls the shift-register via an alternating clock-pulse stage with two complementary outputs A and B.

For each increment the dynamic register comprises a NOR gate and an *RC* delay element. A description of the first three stages will be sufficient to explain the operation of the register. All the remaining stages operate in exactly the same way.

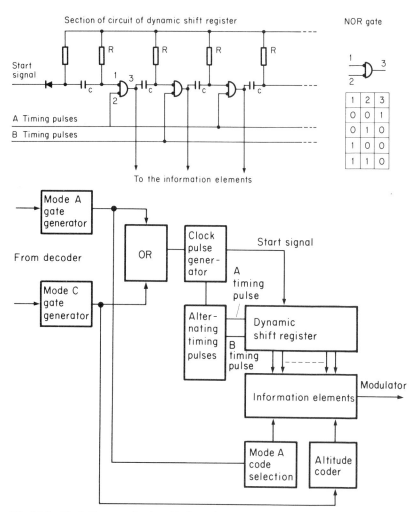

Fig. 3.39 Block diagram of coder with dynamic shift register in ATC transponder AT 6-A

The upper input, 1, of the first NOR gate is maintained in the '1' condition, via a resistor. Any change at the lower input, 2, due to a timing pulse will not cause any change in the output signal. If, now, a mode gate signal appears the clock-pulse timer starts producing its two A and B phases. Simultaneously, a 'start' demand is produced with the result that the level on the 'start' line jumps from 1 to 0. This change is transmitted via a capacitance, that is, as an exponential voltage slope controlled by the time constant of $R$ and $C$.

Thus, the voltage levels at both the upper and lower inputs to the first stage of the register are at 0, with the result that the output level is at 1. This output signal is supplied to element 1 of the information circuits. Cross-talk of the jumping of this signal to the second stage of the register is excluded because the lower input is, at this moment, at 1 due to the B phase-timing pulses. When the next timing phase occurs, the phase A pulse changes to 1 with the result that the output pulse changes from 1 to 0. This change is transmitted via the capacitor C to the upper input of gate 2 as a voltage, rising exponentially from 0 to 1.

Since the B timing pulse is complementary to the A pulse, the lower input of gate 2 is also at '0', so that a 1 signal appears at the output. This 1 signal is then supplied to element 2 of the information circuits. This has no effect on stage 1 since the $RC$ time constant is such that the transmitted voltage again rises to 1 within a period which lies between the 1 and 2 timing pulses. Thus, with a fixed timing of 1.45 μs, sufficient tolerance is provided to produce an exponential function by means of $RC$ elements and to meet the response threshold of the logical gates. It is only therefore the third stage of the register which can accept the negative voltage slope from the second stage, with the consequence that a 1 appears at its output.

For all the remaining stages the information pulse is shifted in exactly the same way. When using integrated semiconductor techniques a dynamic register of this type can be produced with very few components.

### 3.5.6. ATC transponder KT 75

The ATC transponder KT 75 (Fig. 3.40) (made by King Radio Corporation, Olathe, Kansas, U.S.A.) belongs to the lower-priced category. This same manufacturer, using similar design principles, has produced a

Fig. 3.40 ATC transponder KT 75 made by King Radio Corporation

whole series of aircraft radio and navigation equipment in the 'Silver Crown' series, specially designed for the smaller aircraft used in general aviation.

The unit Type KT 75 represents a particularly inexpensive and lightweight transponder. A method of assembly has been chosen whereby the transponder and its control items form a single unit for location on the aircraft instrument panel. This saves the usual cabling between the instrument and its control unit, as well as the filters, required to protect each signal line from interference, which can often be remarkably expensive items.

For special locations the transponder KT 75 can be supplied in an arrangement with a separate control unit, the Type KT 75 R. The cost of this version is about 20% higher than the standard equipment. This is a useful indication of the saving that can be made by combining the transponder and control unit.

There are no special features in the electrical design of the equipment, although the large number of integrated circuits is remarkable for a unit in this price category.

In the transponder KT 75 the controls are arranged to be easily seen. Even the brightness of the reply-monitoring lamp is automatically controlled by the illumination of the environment, to prevent dazzle.

### 3.5.7. ATC transponder TPR-610

This small transponder TPR-610 (Fig. 3.41) (made by Bendix Avionic Division, Fort Lauderdale, Florida, U.S.A.) is designed for location on the instrument panel. All the control units are mounted on a 62 cm$^2$ front panel. There are two further variants of this model to suit different assembly conditions; a flat housing, with the controls in an area of $16.4 \times 3.8$ cm and a 'half-sized' housing with the controls located on a $8.2 \times 6.6$ cm panel.

Apart from the small physical size, the very low current consumption of the TPR-610 must also be mentioned, viz. 0.4 A at 28 V or 0.8 A at 14 V.

In the coder circuit of the TPR-610 (Fig. 3.42), mode gates are first produced by the mode-coincidence signals from the decoder. These mode gates store the interrogation information for the duration of a reply. Each of the two mode gates, for identification and altitude, can set the phase-locked time-pulse oscillator in oscillation via a mixer unit. To produce a serially-

Fig. 3.41 Two models of the ATC transponder TPR-610 with an omni-directional antenna, made by Bendix

pulsed reply train the separate pulses must be produced in sequence. Several digital methods, by which the time delay between the pulse positions can be achieved, have already been explained.

In the transponder TPR-610 provisions have already been taken to provide the unit with just such a delay element; namely, the decoder delay line. In actual fact this can be very simply applied to both tasks. Decoupling between decoding and coding is achieved by the polarity of the input signal. The receiver video signal is supplied to the line with a positive amplitude, and the decoder circuits only respond to positive pulses at the respective tappings of the delay line.

The input signal for coding is a single pulse of negative polarity, and this is read from tappings on the delay line, separated from one another by multiples of 1.45 µs.

Both the framing pulses and the information pulses for identification and height are formed in a diode matrix, the so-called code acceptance gate, by coincidence with these output pulses from the delay line and the information levels set in to the control unit of the altitude coder.

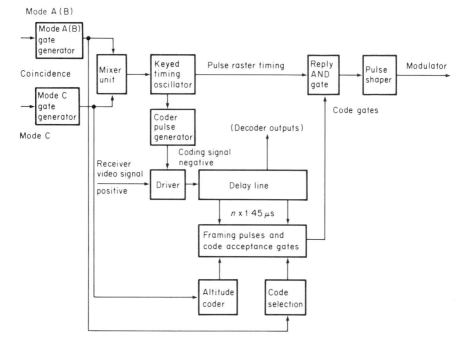

Fig. 3.42 Part of a block diagram of the coder showing the delay line in the ATC transponder TRR-610

The information pulses, however, do not conform in shape or position to the standards of a secondary radar reply. They are therefore better designated as coding-gate signals. These coding-gate signals are then used to synchronize an AND gate in the required pulse positions at the exact pulse-timing produced by the clock-pulse generator. They are thus located exactly in the standard pulse-raster positions, and can then be standardized for pulse-width and amplitude. The pulsed reply train so produced is then applied to the modulator.

# 4. Various Applications of the Secondary Radar Process

Having outlined the basic principles of secondary radar, the units involved and the circuit techniques employed it is worth considering the practical applications. These are both numerous and varied therefore the examples which follow are restricted to basic details. The applications of secondary radar techniques are also continually being extended, consequently only a very general survey can be presented of the current uses and the problems involved.

## 4.1. In the Military Sphere

It has already been mentioned that the present-day secondary radar system employed in civil aviation has been developed from a military system, the original purpose of which was to distinguish friendly and enemy targets on the radar screen. This is still an essential task. Of course, it must at once be obvious that the code capacity of the SIF process is quite insufficient for an exemplary IFF system that is to be free from interference and spoofing. Some assistance can be rendered by additional flying methods correlated with the change of code, and also by frequently changing the code in use.

The application of the secondary radar process to military air traffic control, however, is at least equally important since military and civil aircraft, quite apart from their particular roles, make use of what is essentially the same air space. Only when all aircraft in a common air space are subject to a common system of air traffic control can the safety of the air space be ensured.

A further possible application for secondary radar is the means of distinguishing by different codes several aircraft or groups of aircraft, classified according to their assignments, for purposes of tactical control. The most important military applications employ these techniques but for obvious reasons no detailed description of military methods of operation can be given here.

## 4.2. Civil Air Traffic Control

In civil air traffic control the secondary radar process is of the very greatest importance. Here only a very general review of the current situation can be given. First of all a few terms used need to be explained.

By *navigation* is meant the art of reaching a specified goal from a known starting point along a desired route. There are a number of auxiliary accessories designed for this purpose, some of which are made available by the air traffic control organization. It must, however, be very distinctly kept in mind that there are also navigation devices solely under the control of the aircraft. Assuming that only a single aircraft is to use the air space, then it can do so without any air traffic control purely by means of its own autonomous auxiliary aids.

The concept of *location* implies that the position of a target can be determined in relation to one's own position. Thus, radar and secondary radar are location processes.

The purpose of *air traffic control* is to ensure that several aircraft can simultaneously use the same air space with the greatest possible safety, and to enable each individual user to select his time and flying route without unreasonable limitations.

In such a system of air traffic control the aircraft with their pilots, the airports, navigation systems, aeronautical information services, meteorological information services, flight regulations and flight procedures, the actual air traffic control equipment, the telecommunications links, the air space monitoring equipment and many other ground installations all play a necessary part.

The scope of and mutual interactions between these separate services differ widely, depending on traffic density, geographical and political factors and, by no means least, on economic considerations. The historic development and the present state of this technique also play a definite part.

The spectrum of air traffic situations thus extends from those sparsely-populated areas where air traffic control can be limited to the provision of a few simple navigational aids, and the observation of the basic traffic regulations up to those focal points for air traffic, such as the air space between Washington and New York where today, despite the latest technical aids, it is still almost impossible to avoid delays.

Air traffic is controlled in accordance with two basic sets of regulations: the visual flight rules (VFR), and the instrument flight rules (IFR).

The visual flight rules (VFR) are based on the assumption that each air space user can see any other air space user in plenty of time to avoid him, by observing fixed traffic rules just as in the instance of road traffic. The basic assumption for the performance of any VFR flight is, therefore, proper visual conditions.

On an IFR flight: under instrument flight rules, that is: visibility during the flight is not necessary. Before commencing the flight the pilot makes out a flight plan on which he states the route he will follow by means of his navigational instruments, and at what time this will probably take place. In most instances the route will follow flight lanes, defined by navigational aids, or otherwise-controlled air spaces. The air traffic control authorities, by coordinating the various flight plans, endeavour to ensure that the separation between each aircraft does not fall below certain safety limits.

To check that the actual flight corresponds with the flight plan it is also necessary to monitor the air space. In earlier days this monitoring of the air space was done in accordance with the so-called rules of procedure. The already-declared flight plan had to be accurately effected section by section, and had to be officially acknowledged by reporting one's position over the radio telephone when flying over specified reporting points. By and large the process is similar to that in a big warehouse, where the outgoing and incoming goods are checked by a clerk. This process can be partially performed by data processing equipment. Such modern methods will ease the staff problem and will help to increase reliability.

This process is, however, limited by the fact that the position of an aircraft between two reporting points can only be extrapolated from its previous behaviour. This is, accordingly, inaccurate and requires that large safety margins must also be taken into consideration.

The rules of procedure do not facilitate really satisfactory use of the air space. Consequently, with the increase in the amount of traffic, efforts have been made to use systems of air space monitoring by means of radar processes. Initially, it was only possible to do this to the extent of identifying aircraft by reports of their positions, flying procedural turns or other not entirely reliable measures and to follow them continuously on the primary radar screen. Only by the introduction of the secondary surveillance radar system was it possible to identify every aircraft equipped with a transponder, and to reduce the separation standards by continuous monitoring from the ground. So that the radar observer may know those aircraft in which he may expect to find transponders, the flight plan indicates an aircraft equipped with a transponder by means of an alphabetical code.

The allotment of the reply codes is still treated in many different ways.

A method used in the U.S.A. is to allot the code according to the type of flight, approximately as shown in Table 12.

Table 12   Interrogation Mode A

| | | Code |
|---|---|---|
| VFR flights | Below flight level FL 33 | 3300 |
| VFR flights | FL 100 and above up to the lower limit for positive air space control (in U.S.A. FL 180 or FL 240 depending on geographical position). | 3400 |
| VFR flights | Above FL 240 (if positive air space control is not available). | 3500 |
| IFR flights | On take-off, climbing to a crusing altitude below FL 240. | 1000 |
| | Take-off between ground and FL 240 | 1100 |
| | Climbing to FL 240 and above. | 2000 |
| | FL 240 and above. | 2100 |
| *En route* flights on IFR | Below FL 240 | 1100 |
| | FL 240 and above | 2100 |
| | Sudden change of altitude by more than one level. | 0500 |
| Arrival code | After receiving permission to descend. | 1400 or 0400 |
| IFR | Waiting at FL 240 and above. | 1100 or 2100 |
| | Transfer to airport control. | 0400 |
| Emergency | General | 7700 |
| | Communications failure | 7600 |
| | Hijacking of aircraft | 3100 |

Table 12 shows that the secondary radar process, besides being of importance for IFR flights in accordance with instrument flying regulations, is also used to identify VFR flights according to visual flight rules. Both groups, IFR and VFR flights, are taking place in the same air space so that it is axiomatic that the reliability of the air traffic control will depend on the numbers of aircraft it can monitor.

Other codes can be allotted to distinguish all aircraft which will be under the control of any specified radar observer. Proposals have also been made to allot serial numbers to all commercial flights and thus introduce individual identification.

The application of altitude information has been illustrated in detail in Section 1.2.

The possibilities of combining different codes with special signals such as SPI, general emergency and the indication of a radio failure can also be found in Section 1.2.

In addition to identification and transmission of altitude data SSR can be used to enhance primary radar returns and in certain special circumstances be used on its own to provide positional information.

Within the normal covering range of primary radar installations it is common to have complete gaps or areas of low probability of detection. These are caused by meteorological effects and variations in the conditions of propagation. In such instances secondary radar positional data is a valuable supplement which enhances the value of the primary radar data tor ATC purposes.

It can be shown that a positive system of radar surveillance for ATC has only been made possible by the use of secondary radar. The operational use of this facility has not yet been standardized and its use in the future will also depend upon the operational environment in which the system is required to be used.

# 5. Some Technical Expressions used in Secondary Radar Techniques

*A-pulses.* Those pulses in SSR reply which occur 2.9, 5.8 and 8.7 µs after the first framing pulse (F1).

*Acceptance Gate.* A gated signal produced in the defruiter (q.v.) by the video signal stored from the previous recurrence period. This gate signal allows acceptable incoming video to be passed out of the defruiter.

*Acceptance Probability.* The probability of the acceptance of a reply pulse when correlated with an acceptance gate in a coincidence circuit in the defruiter.

*Active Decoder.* A device used in an SSR ground unit. After selecting the coordinates of an aircraft, the information content of the SSR replies transmitted by that aircraft can be decoded and displayed numerically.

*Active Readout.* An apparatus for the numerical display of actively decoded secondary radar responses.

*AIMS.* Program name for IFF-system in U.S.A. Acronym of *A*TCRBS *I*FF *M*ark XII-*S*ystem.

*Airport Surveillance Radar (ASR).* A short-range radar for air traffic control as laid down in the FAA regulations.

*Air-route Surveillance Radar (ARSR).* A long-range radar system for air traffic control on air lanes as laid down in the FAA regulations.

*Air Traffic Control Radar Beacon System (ATCRBS).* A civil secondary radar system for air traffic control in accordance with the FAA regulations.

*All A/C.* All aircraft (q.v.).

*All Aircraft.* An operational function in SSR ground units. In this method of operation all transponding aircraft within the interrogation range are displayed on the PPI. The targets, which respond with Mark X.SIF framing pulses are displayed irrespective of the code content of the reply. Mark X Basic responses are fully displayed.

*All Codes (AC).* An operational selection in SSR ground equipments. In this method of operation all transponding aircraft within the interrogation range are displayed on the PPI independently of the code content of the reply.

They are indicated by a single line (Positional Response, q.v.). Aircraft replying in the basic Mark X-IFF system are not displayed. The display can be limited to those aircraft that reply to a particular mode. This latter choice is called Select Mode/All codes.

*All C/S.* All Common System, the same as *All Codes.*

*Alphanumeric Display.* The display of SSR information related to an aircraft in alphabetical letters, numerical figures and symbols on PPIs or on separate alphanumeric indicators.

*Altitude Decoder.* A decoding device for translating the altitude code transmitted in reply to a mode C interrogation. It translates the MoA-Gilham code, used for encoding the aircraft reply into a three-digit decimal display.

*Altitude Digitizer.* An altitude-measuring system with a digital output which converts the pressure altitude into the MoA-Gilham code for encoding the aircraft reply. It may also be called Altitude Encoder.

*Altitude Filtering.* Only those aircraft are selected for display which are within a specified altitude layer (zone). The same phrase is given to the process where those aircraft between two heights are identified by an extra bar or special symbol on the PPI.

*AOC.* Automatic overload control (q.v.).

*ARSR.* Air-route surveillance radar (q.v.).

*ASR.* Airport surveillance radar (q.v.).

*ATCRBS.* Air traffic control radar beacon system (q.v.).

*Automatic Overload Control (AOC).* A device in the transponder which limits the reply rate to prevent overloading of the transmitter tubes.

*B-pulses.* Those code pulses in an SSR response which occur 11.6, 14.5 and 17.4 μs after the first framing pulse, F1.

*Back Lobe.* The radiation pattern in the horizontal polar diagram for a directional antenna which is directly opposite that of the main lobe.

*Bar.* A bar-shaped mark for an aircraft on the PPI to indicate that the target is replying with the same SSR reply as selected on the code setting unit – be it all code, select code, etc.

*Basic Mark X-IFF System.* A military identification system with three modes of interrogation, the predecessor of the currently used Mark X-SIF system. Some transponders of this type are still in use. Compared with the later system they have a much more restricted scope for coding the reply.

*Beacon.* An electronic device which automatically transmits a reply signal if it receives a specified interrogation signal.

*Beacon Assist.* A means of distinguishing all the aircraft on the PPI which reply to any interrogation mode, with a preference for those which transmit a specific code.

*Beacon Video.* The video output of an SSR interrogator receiver consisting of aircraft replies often preceded by mode pulse pairs at the start of the recurrence period.

*Beacon Video Digitizer.* A device for the automatic processing of all the SSR replies received.

*Beacon Sync Trigger.* See under Pre-trigger.

*Beam Sharpening.* A reduction in the effective width of the main beam of an SSR aerial by using one of the side-lobe suppression processes, or by using the mono-pulse antenna technique.

*Beam Width Control.* A process by which the width of the arc of the SSR reply can be adjusted on the radar screen so that, for example, it is dependent on range (see Beam Sharpening).

*BITE.* Built-in test equipment (q.v.).

*Blip.* A target-echo point on a PPI.

*Bloomer.* An SSR reply marking on the PPI which is expanded radially for special circumstances, e.g. emergency replies.

*Blooming.* Thickening of the light spot (see Bloomer).

*Brackets.* Framing pulses (see Bracket Pulses).

*Bracket Decoding.* A process in decoders for identifying valid responses which makes use of the fixed interval of 20.3 μs between brackets (framing pulses). At the same time this process supplies range information (see also Positional Response). Any code content of the SSR reply is ignored in this process.

*Bracket Pulses.* These are the marker pulses of an IFF Mk X-SIF reply. They are spaced 20.3 μs apart (leading-edge to leading-edge) F1 and F2.

*Built-in Test Equipment (BITE)*. Automatic test equipment which generally performs 'go-no-go' tests.

*BVD*. Beacon video digitizer (q.v.).

*C-Pulses*. Those pulses in an SSR reply which occur 1.45, 4.35 and 7.25 µs after the first framing pulse, F1.

*Caboose Pulse*. A slang expression for a special position identification (SPI) pulse (q.v.).

*Capture (SSR)*. The term given to that property of transponders by which they only respond to the interrogations from one ground station in preference to another for one of several reasons, e.g. receiver desensitization due to reply overload or due to transponder dead time.

*Centre mark*. A mark, indicating the centre of the target.

*Centre marking*. Indicating the centre of the target, for example, by means of a radial stroke on the SSR target arc.

*Challenge*. A call, interrogation, or demand.

*Code*. See Reply Code.

*Code Garbling*. See Garbling.

*Code Response*. The two arcs in a 'double-slash presentation' (q.v.) which, in the American ATCRBS system and in the English SSR system, indicate that the target is replying with a specified code. (See also Positional Response, for comparison).

*Code Train*. The pulses in an SSR reply between the bracket pulses.

*Coder*. A device in a secondary radar installation which produces the required series of pulses for the modulation of the transmitter. The coder within the transponder produces the series of pulses associated with the required reply code, whilst the coder in the interrogator unit produces the series of pulses for the required interrogation mode.

*Coincident Display*. The SSR replies are usually displayed at a time (or range) delay on primary radar in order to separate the information. Where however an arrangement is made to display SSR replies at the time range this is known as 'coincident display'.

*Common Decoder*. A central equipment SSR ground installations which, amongst other tasks, performs a series–parallel conversion of the SSR reply code, and ensures that incoming video are genuine aircraft replies.

*Common System (C/S).* A common air traffic control system used both by civil and military aircraft with particular reference to the connection between the ATCRBS and IFF Mark X-SIF systems.

*Communication Failure.* A failure in the radio speech link. To indicate this an emergency signal in code 76 or 7600 is transmitted in reply to interrogations in mode 3/A or B. The actual reply code is identical for either 76 or 7600. These numbers represent the code selection made in the aircraft on the old and new type of transponders respectively.

*Control Pattern.* The radiation pattern of the antenna used for transmitting the $P_2$ pulse.

*Control Pulse.* A pulse ($P_2$) emitted by the SSR ground equipment for sidelobe suppression (q.v.).

*Correlation Criteria.* The acceptance gate created by an SSR reply exists for a number of recurrence periods during which time a reply or more than one reply from the same range must have been received before the replies are considered acceptable.

*C/S.* See Common System.

*Count Down.* The ratio of the number of valid interrogations received by the transponder to the number of replies emitted by the aircraft.

*D-pulses.* Those pulses in an SSR reply which occur 13.05, 15.95 and 18.85 μs after the first framing pulse $F_1$.

*Data Link.* A data transmission system specially used for air traffic control purposes between aircraft and the ground station.

*Data Transmission Feature.* A technique for the transmission within the response code of information additional to identification. Such additional information, for instance, as data on altitude, flight path and relative speed of the aircraft.

*Dead Time (DT).* See Transponder Dead Time.

*Decoder.* A device in SSR ground installations for decoding the information contained in the aircraft reply and in the aircraft SSR-transponder for decoding the mode of interrogation.

*Defruiter.* A synchronous filter device for the suppression of unsynchronized responses. (See also Defruiting.)

*Defruiter Criteria.* These indicate how the fruit or non-synchronous replies are filtered out, i.e. after how many storage periods, and in what combinations, e.g. for '2 out of 2' see Single Defruiting and for '3 out of 3' see Double Defruiting.

*Defruiting.* A process by which all the aircraft replies accepted by the interrogator are tested by means of storage and a comparator for synchronism with the interrogation repetition frequency. Only those replies which are in synchronism pass through the filter.

*Defruiting Efficiency* $\eta = 1 - (n_a/n_e)$. Where $n_a$, and $n_e$ are respectively the number of 'fruit' responses at the output and input of the defruiter.

*Degarbling Circuits.* Supplementary equipment in SSR decoders which detects and in specified instances suppresses the decoding of the SSR reply if (see Synchronous Reply-code Overlap) pulses appear within the normal reply code train so as to prevent decoding with the usual degree of confidence.

*Difference Pattern.* The receiver characteristic of a monopulse antenna obtained by connecting together in anti-phase the signals received by the two partial antennae. The difference pattern has a minimum in the main radiation direction of the antenna.

*Digitizer.* An analogue-to-digital converter.

*Displayed Beam Width.* Breadth of the arc of the SSR reply displayed on PPI.

*Ditch.* The negative-bias voltage produced by the echo-suppression circuit in the transponder IF amplifier.

*Double Defruiting.* A method of filtering responses which, statistically, are not synchronously distributed, in which three sets of video signals from the same range over three or more interrogation periods are compared with one another, e.g. the '3 out of 3' criterion.

*Double-slash Presentation.* The marking of an aircraft on the PPI to indicate, in the American ATCRBS and English SSR systems, that the target is responding with a specified code (see also Positional Response and Code Response).

*DT.* See Dead Time and Transponder Dead Time.

*DTF.* Data transmission feature (q.v.).

*Echo Suppression.* A transponder circuit which, for a short period, reduces its receiver sensitivity. This reduction in sensitivity makes the transponder less liable to reply to a delayed version of an interrogation signal as a result of a reflection.

*Effective Beamwidth.* A sector in the directional characteristic of a ground antenna within which aircraft responses are triggered and received.

*Emergency.* A special response from the aircraft transponder to indicate that the aircraft is in an emergency situation. This emergency response may comprise an SSR reply repeated four times in succession (a response code followed by three bracket pulses – pairs of framing pulses), or a special code (Civil code 77, or 76, in the event of a failure in the radio speech link), or four pulses at intervals of 16 μs (in the basic Mark X-IFF system).

*Environment.* The presence of a number of SSR ground stations operating in a particular area materially affects the performance of all the ground stations and also all the aircraft transponders in that area. This interaction creates the SSR environment of a particular area.

*External Suppression.* The transponder is suppressed by a signal which is produced whenever another unit in the aircraft transmits on a frequency that affects the transponder.

*F1, F2.* Framing pulses (q.v.).

*False Code.* A reply from a transponder, where the content has been changed by interference pulses detected in the interrogator.

*False Garbling.* A simulated 'garbling' situation where an interference pulse either precedes or follows an aircraft reply such that the interval between it and a code pulse is 20.3 μs.

*False Moding.* A transponder replies to a different mode from the mode on which it was interrogated.

*Filtered Display.* The selection of specified SSR replies for display on the PPI, especially with the aid of the so-called passive decoder, the selection being by identity or altitude, for example.

*Folded Pillbox Antenna.* The designation for a special type of antenna with a very weak side-lobe formation.

*Framing Pulses.* See Bracket Pulses.

*Fruit.* Irregular SSR replies, received by an interrogator, that were initiated by other interrogators at different p.r.fs.

*Fruit Density.* The number of 'fruit' replies per unit time (usually, per second).

*Gain-time Control (GTC).* A device in the receiver of the ground unit which effects control of its sensitivity dependent on time and, consequently, on range, to prevent the reception of aircraft replies via the side and back lobes of the ground antenna, and to reduce the range of amplitudes of aircraft signals received through the main beam.

*Garble Sensing.* A process in SSR ground installations for detecting garbling and for decoding the resulting codes as far as possible.

*Garbling.* A process whereby two or more reply signals overlap in time after detection in the interrogator with the result that the information in the reply can be made up from more than one aircraft. (See also Non-synchronous Reply Code Overlap and Synchronous Reply Code Overlap.)

*GAT.* General aviation transponder.

*Gate Circuit.* A logic element with $n$ inputs and a single output by which other circuits can be switched on or off, either by energizing at least one input (an OR gate) or only by energizing all the inputs (an AND gate).

*Ghost.* A false target indication due to the fact that the interval between the reception of two response signals is such that pulses of one response are separated from the pulses of the other response by 20 µs, i.e. framing pulse spacing.

*Go-no-go Test.* A method of testing which provides results in the form of a 'Yes' or 'No' statement.

*GTC.* See Gain-time Control.

*Height Layering.* See Altitude Filtering.

*Hit.* An interrogation signal which reaches the transponder antenna and which exceeds the threshold of the receiver. (See also Reply Hit.)

*IDENT.* A special response code from an aircraft which is switched on when the pilot operates the microphone switch. 4.35 µs after the second framing pulse, in a normal reply, a special (SPI) pulse is added to distinguish one aircraft from others. An IDENT response is displayed on the PPI in the form of a wide beam marker or, in some installations, in the form of a double bar. (See also Caboose Pulse and SPI.)

*Identification Friend or Foe (IFF).* A method of aircraft identification by secondary radar.

*Identification of Position (I/P)*. A means of distinguishing one aircraft from several others by the single repetition of the reply or of a pair of framing pulses 4.35 µs after the second bracket pulse. This signal is triggered by pressing the I/P button or by operating the microphone button. This is the military equivalent of the IDENT function (q.v.).

*IFF*. Identification Friend or Foe (q.v.).

*Identification Response*. See Code Response.

*IFF Mark X-SIF System*. The military version of the ATCRBS system. This installation corresponds to the Mark X-IFF system with the addition of a coder and decoder which are used to extend the coding scope of the aircraft reply.

*Information Pulses*. The pulses containing the message, which appear between the two framing pulses in an aircraft reply. They are spaced at 1.45 µs intervals and, depending on their position, are defined as A, B, C, or D pulses.

*Interference Blanker*. The same as a defruiter (q.v.).

*Interlace*. See Mode Interlace.

*Interlace Ratio*. The ratio of the number of interrogations per second in different modes when interrogating on several modes. (See also Mode Interlace.)

*Interleave*. Two series of reply trains become superimposed in time in such a way that their time spacings can be distinguished and separated out. (See also Garbling.)

*Interrogation*. The process in which an SSR ground unit transmits coded pairs of pulses through a directional antenna to trigger responses from airborne units. (See also Mode and Challenge.)

*Interrogation Beam Width*. The angular sector in the directional characteristic of an SSR ground antenna within which responses from aircraft are triggered.

*Interrogation Mode*. See under Mode.

*Interrogation Path*. The transmission path during an interrogation.

*Interrogation Path Side-lobe Suppression (ISLS)*. A method of preventing replies from aircraft to interrogations received at the transponder from the side lobes of the ground antenna. This method involves a comparison, within the transponder, of the amplitudes of the pairs of pulses in the interrogation with a control pulse also emitted by the ground station. This control pulse can be identical with one of the pulses in the pair of interrogation pulses which is radiated antidirectionally, in addition to the main beam.

*Interrogation Pulse Pair.* The two pulses, $P_1$ and $P_3$, which are the constituents of every SSR interrogation. See also Mode.

*Interrogation Recurrence Frequency (IRF).* The number of pairs of pulses produced per second by an SSR interrogator.

*Interrogator.* The part of an SSR ground installation in which the RF power is produced and is modulated by the mode pulses. This expression is also used as the abbreviation for an interrogator responsor (q.v.).

*Interrogator Responsor (I/R).* An SSR ground unit which transmits interrogations on one frequency and receives aircraft replies on another.

*I/P.* Identification of Position (q.v.).

*I/R.* Interrogator Responsor (q.v.).

*IRF.* Interrogation Recurrence Frequency (q.v.).

*ISLS.* Interrogation Path Side-lobe Suppression (q.v.).

*Kill.* In its stricter sense this means disabling circuits. This expression is used to indicate operation of circuits in the ground-unit decoder which suppress the secondary radar reply if this has been garbled by another response signal (see Degarbling). It is also used to denote the operation of the side-lobe suppression circuit in the transponder which prevents the transponder replying to interrogations from the side lobes of the interrogating antenna.

*Killer Circuits.* Logical circuits in the transponder which prevent the transmission of replies to the interrogations from the side lobes of the interrogating antennae that have ISLS.

*Light Gun.* An opto-electronic device for selecting aircraft for active decoding.

*Light Pen.* An opto-electronic device for selecting aircraft for active decoding.

*Low Sensitivity.* A transponder control. The transponder sensitivity is reduced below its normal value by actuating a switch. This reduces the effect of side lobes in the neighbourhood of the interrogator unit.

*Main Beam Killing.* The failure to elicit a reply within the coverage range of the main beam of the ground antenna from a transponder. This is a consequence of the limited dynamic range of the transponder receiver in cooperation with the ISLS circuits, and the phenomenon only happens during the triple pulse process.

*Mark X-IFF.* See Basic Mark X-IFF.

*Mark X-SIF.* See IFF Mark X-SIF System.

*Mark XII.* A military identification system.

*Minimum Triggering Level.* The minimum power level at the input to the receiver for valid pairs of interrogator pulses to permit aircraft replies to be triggered with a minimum probability of 90%.

*MoA-Gilham Code.* A system of coding aircraft altitude recommended by ICAO for replies to interrogations in Mode C. In this code a change of one unit of measurement in the value of the altitude to be transmitted only alters the coded information by one bit.

*Mode.* Description of the coding of an interrogation. The interrogations consist of two pulses, $P_1$ and $P_3$, with the following intervals:

| | |
|---|---|
| Mode 1 | 3 μs |
| Mode 2 | 5 μs |
| Mode 3/A | 8 μs |
| Mode B | 17 μs |
| Mode C (transmission of altitude) | 21 μs |
| Mode D (as yet unused) | 25 μs |

*Mode Gate.* A signal which indicates for the duration of a single interrogation period in what mode the interrogation is being made.

*Mode Interlace.* A series of interrogations made in two or more modes.

*Monopulse.* A process by which the effective width of the main beam of an antenna is reduced by special processing of the received signal.

*Monopulse Decoder.* A decoder which only relays to the indicator unit the first of the series of pulses in a reply. It is used for the display of all aircraft equipped with a Mark X-SIF or ATCRBS transponder.

*MTL.* Minimum Triggering Level (q.v.).

*Multiple Decoder.* A decoder which permits several (four or ten) coder responses to be evaluated simultaneously.

*Non-common Decoder.* A device sometimes associated with a PPI for the passive decoding of aircraft replies.

*Non-synchronous Reply Code Overlap.* Overlapping of the reply trains of two or more transponders at the interrogator, which do *not* have a common time scan.

*Normal Sensitivity.* A sensitivity, of approximately $-75$ dBm, to which transponders are set under normal operating conditions.

*Omni-notch Characteristic.* Antennae with this type of characteristic are frequently used to transmit ISLS pulses. (See also SLS and Control Pulse).

*Over Interrogation.* Interference in the operation of a secondary radar system due to the fact that the excessive number of interrogations exceeds the capacity of the transponder.

$P_1$, $P_3$. The two pulses comprising the pair of pulses in any interrogation mode. (See also Mode.)

$P_2$. Control Pulse (q.v.).

*Parrot.* Slang expression for a transponder.

*Passive Decoder.* An electronic device in SSR ground equipment which only supplies a pulse for display on the PPI after the code content in the aircraft reply has been checked for correlation with a code set on a control unit.

*Phantom.* See Ghost.

*Plan Position Indicator (PPI).* A video display unit for the display of decoded and undecoded video signals in polar coordinates.

*Positional Response.* In the 'select' mode of operation (see Select) this is the first of the two target arcs to indicate a target aircraft responding in the correct code. It indicates the position of the replying aircraft. (See also Code Response.)

*Positional Signal.* See Positional Response.

*PPI.* See Plan Position Indicator.

*Pre-trigger.* A trigger pulse which synchronizes the SSR ground equipment with the primary radar so as to ensure that the aircraft replies are properly timed in relation to the primary echo.

*PRF.* Pulse Repetition Frequency.

*Primary Radar.* A radar unit which does not make use of active radiation on the return path.

*Range Bin.* An increment in range – the range coverage of a radar installation being divided into discrete steps.

*Range Box.* See Range Bin.

*Rapid Read-out.* See Active Decoder.

*Raw Video.* Unevaluated video information used for display.

*Receive Subtraction.* See Reply Path Side-lobe Suppression.

*Receiving Path Side-lobe Suppression (RSLS).* The same as Reply Path Side-lobe Suppression (q.v.).

*Recovery Time.* The time within which the transponder sensitivity rises to 3 dB below its normal value after its reduction for purposes of echo suppression.

*Reflections.* Unwanted signals, due to interrogations or replies being reflected from ground obstacles such as aircraft hangars, buildings, metal towers or hills.

*Rejection Probability.* The probability that a reply will be rejected in the defruiter (the rejection probability is the complement of the acceptance probability).

*Reply Code.* A digital-coded pulse train used for the transmission of information from the aircraft to the ground. It comprises two framing pulses $F_1$ and $F_2$ at an interval of 20.3 μs within which up to 12 pulses can be inserted. (These code pulses are spaced at 1.45 μs.)

*Reply Hit.* A response from an aircraft, triggered by a valid interrogation, which is received on the ground.

*Reply Hits Per Scan.* The number of reply hits received from a target per revolution of the ground antenna.

*Reply Path Side-lobe Suppression.* A method of suppressing the responses received via the side lobes by the ground antenna.

*Responder.* The response transmitter.

*Responsor.* The receiver of responses.

*Reserve Gain.* The signal strength of the interrogation or the reply at a point in space greater than the path loss to the receiver.

*Ring Around.* A typical circular pattern on the PPI of SSR replies caused by continuous reception of aircraft replies to interrogations by the side lobes of the ground antenna. Normally this only occurs at very short ranges.

*Round Reliability.* The ratio of the number of valid replies detected to the number of valid interrogations transmitted.

*RSLS.* See Receiving/Reply Path Side-lobe Suppression.

*Second Train.* This is used for purposes of location or special identification (see I/P).

*Secondary Radar.* In contrast with the traditional radar process, which depends on the processing of the radar transmitter signal after its partial reflection (echo) from an aircraft target, in the secondary radar process a response signal, generally coded, is triggered in a responder unit by the signal from an interrogator station. The responder may operate at the same or at a different frequency from the interrogator.

*Secondary Surveillance Radar (SSR).* A method standardized by ICAO whereby information (identification, altitude) can be transmitted from an aircraft to the ground, in addition to location, by aircraft equipped with the appropriate aircraft unit (see Transponder).

*Select.* A control function in SSR ground units. In this mode of operation the replies received by the secondary radar are passively decoded. (See Passive Decoding. For comparison, see also All A/C and All C/S.)

*Select Aircraft.* See Select.

*Selective Identification Feature (SIF).* A system of coding for replies in the Mark X identification system.

*Sensitivity Time-control (STC).* Similar to Gain Time Control (q.v.).

*Setrin SLS.* A method of side-lobe suppression proposed by M. Setrin of RADC. In this method a triple pulse interrogation is used. A control pulse follows the first interrogation pulse at an interval of 2 μs. (See also Three-pulse Side-lobe Suppression.)

*Shrimp Boats.* Small discs of transparent material on which the radar operator makes brief notes with a wax pencil. These discs are pushed along in the proximity of the target on horizontal radar screens.

*Side Lobe.* Radiation in an undesirable direction in the radiation pattern of the SSR ground antennae.

*Side-lobe Suppression (SLS).* The term for the process in SSR whereby the consequences of side lobes in the directional characteristic of the ground antenna are eliminated.

*SIF.* See Selective Identification Feature.

*Single Defruiting.* In this process two SSR replies from the same point in space in two or more interrogation periods are required to produce an output from a defruiter.

*Slash.* A spot of light, produced on the PPI by a secondary radar response, distinguished by the fact that its dimension in azimuth is greater than its radial dimension.

*SLS.* See Side-lobe Suppression.

*Special Position Identification Pulse (SPI).* A pulse, occurring 4.35 µs after the second framing pulse, in the response from a transponder. It is triggered briefly by the pilot of the aircraft at the request of the ground station and produces a special indication of the target on the PPI. (See also Caboose Pulse and Ident.)

*SPI.* See Special Position Identification Pulse.

*Spike Discriminator.* A noise spike rejection circuit.

*Split.* The display of an SSR target on the PPI where the slash is not continuous. The gaps can be caused by several means.

*Squawk.* Slang term for a secondary radar reply.

*Squawk 2, Squawk 3, Squawk Alpha, Code. . . .* A demand for the pilot to switch on Mode 2, Mode 3, or Mode A and Code. . . .

*Squawk Flash.* A demand for the pilot to press the I/P button.

*Squawk Low.* A demand to the pilot to reduce sensitivity, or a demand for low sensitivity (q.v.).

*Squawk Mayday.* A demand to the pilot to switch on the emergency code 7700.

*Squawk Normal.* A demand to the pilot to select normal sensitivity.

*Squitter.* Unwanted and uncontrollable triggering of aircraft responses by interference pulses or by noise.

*SR.* Secondary Radar (q.v.).

*SSR.* Secondary Surveillance Radar (q.v.).

*Standby.* A switch position on the transponder control unit. Although the transponder, in this position, is not in operation it can be immediately set in operation without the need for any warming up.

*STC.* Sensitivity Time Control (q.v.).

*Storage Tube Defruiter.* A unit for suppressing non-synchronized responses by means of a storage tube (see Defruiter).

*Strangle 2, Strangle 3.* Slang expressions for a demand to the pilot to switch off Mode 2 or Mode 3.

*Strangle Flash.* Slang term for the demand to switch off the I/P.

*Suicide.* The unwanted suppression of a response to a mode of interrogation in a military transponder by the side-lobe suppression circuit.

*Sum Pattern.* The radiation characteristic of a monopulse antenna obtained by combining the signals transmitted or received in the same phase from the two partial antennae. The sum pattern has a maximum in the antenna's main direction of radiation.

*Synchronous Reply-code Overlap.* Overlapping of the reply trains from two or more transponders replying to the same interrogation, having a common time scan.

*Target Killing.* The suppression of a target.

*Three-pulse Side-lobe Suppression.* A control pulse, between the two interrogator pulses, with a field strength slightly higher than that of the maximum side lobe in the directional antenna is radiated by an omni-directional antenna. If the aircraft transponder receives the control pulse at the same or at a higher field strength than the interrogator pulse, the reply is suppressed. (See also Side-lobe Suppression.)

*Threshold.* A level which a signal must exceed if it is to be accepted.

*Transponder.* A unit which transmits a response signal on receiving an interrogation. The expression is a 'portmanteau' word for the terms transmitter and responder.

*Transponder Dead Time.* A period of time after the receipt of a detected interrogation during which the transponder is inhibited from receiving further interrogations.

*Transponder Reply Limit.* The same as Automatic Overload Control (q.v.).

*Trigger Level.* A threshold value which any received interrogation must exceed if it is to trigger a response in the transponder.

*Two-pulse Side-lobe Suppression.* The first pulse in a pair of interrogation pulses is radiated by an omni-directional antenna with the same field strength as the second pulse emitted by the main lobe of the directional antenna. The transponder only provides a response if both pulses are received with the same field strength. (See also Side-lobe Suppression.)

*Video Interconnection Equipment (VIE).* These are decoder units fitted in ground stations in accordance with FAA regulations which decode secondary radar responses and distribute the associated video signals to the display units.

# References

1. F. Trenkle, *Bordfunkgeräte der deutschen Luftwaffe 1939–1945*. Ausschuß für Funkortung (jetzt DGON) Düsseldorf, Best.-Nr. 1031.
2. F. Trenkle, *Deutsche Ortungs- und Navigations-Anlagen (Land und See 1935 bis 1945)*. Deutsche Gesellschaft f. Ortung und Navigation, Düsseldorf, Neudruck 1966, Best.-Nr. 1038.
3. *Secondary Radar* (IFF). Part I. Early History. Part II. Contributions by Frederick C. Williams. IEEE Transactions on aerospace and electronic systems, September 1973, pp. 790.
4. *International Standards and Recommended Practices, Aeronautical Telecommunications: Annex 10 to the Convention on International Civil Aviation (ICAO)*. 2nd Edition of Volume I, April 1968.
5. M. Kayton and W. Fried, *Aviation Navigation Systems*. John Wiley and Sons, Inc., New York-London-Sydney-Toronto, 1969.
6. M. Skolnik, *Radar Handbook,* McGraw-Hill Book Company, New York, 1970.
7. *DIN 40700: Schaltzeichen Digitale Informationsverarbeitung*. Beuth-Vertrieb GmbH, Berlin und Köln.
8. *CAA Paper 76010: ADSEL – A Selectively Addressed Secondary Surveillance Radar System*. Civil Aviation Authority, London.
9. *Technical Development Plan for a Discrete Address Beacon System,* Report No. FAA-RD-71-49 Department of Transportation Federal Aviation Administration, Office of Systems Engineering, Management Systems Research and Development Service, Washington, D.C. 20590.

# Subject Index

A-pulses 196
Absolute height 25
A. C. Cossor Ltd. 174
Acceptance characteristic 57, 115
Acceptance gate 54, 113, 123, 169, 196
   width 56, 113
Acceptance probability 196
Acceptance tolerance 171, 176
   minimum 177
Activation 120
Active altitude decoder 89
Active decoder 84, 123, 124, 125, 196
   control unit 125
Active decoding 32, 89, 123, 125
Active readout 200
Activity display 123
Address code 74
'Advance period' 34
AF interference 137
AIMS 138, 139
Air space 193
   monitoring 193
      equipment 192
   use 193
Air traffic control 173, 191
   authorities 193
Air traffic density 164, 192
Aircraft instruments, minimum requirements 133, 134, 173
Aircraft mains supply 155
Alarm unit 118, 131
'All Aircraft' 116, 119
'All correct' 159
Alphabetical letter code for radar frequency bands 19
Altitude
   decoder 84, 197
   digitizer 197, 202
   error, relative 25
   filtering 24, 197, 203
   information 132, 194
   interrogation 206
   transmission of 24, 27, 195
'Amplitude compression' 105

Amplitude filter 128
AN/APX-46 139
AN/APX-90 163, 165
Analogue delay line 167
Analogue delay unit 181
Angular correlation 34, 43
Annex 10 84
Antenna 79
   blade-shaped 38
   -diagram 37
   diversity operation 165
   drive 79
   flush-mounted annular slot 38
   gain 36
   lead 161
   rotation 68
   sweep 68
   target dwelling time 63, 69, 70
Antennae, problems 35
Anti-suicide 202
AOC 66, 146, 150, 197
   signal 146
APD 12 121–128
ARC Aircraft Radio Corporation 176
Area gating control 123, 125
ARINC 134
   specification 173
Arrangement of units 94, 139, 167
Arrival code 194
ARSR 196, 197
Aspect 65
ASR 196, 197
AT6-A 185
ATC-transponder 173, 174, 176, 179, 183, 187, 188
ATCRBS 12, 196, 197
Atmosphere, actual 25
   correcting processes 25
   standard 25
Atmospheric pressure, low 143, 154
ATR Housing, standard 179
Attenuation 126
   atmospheric 20
Attenuator 128

Automatic decoder  89
Automatic decoding  33
Automatic overload control  46, 66, 197
Automatic test equipment  93
Autonomous navigation  192
Auto-trigger  97

B-pulses  197
Back lobe  197
Backward radiation  41
Bandpass  143
Bandwidth  146
Bar (display)  123
Basic Mark X  119
Basic Mark X-IFF System  198
Beacon  198
   Sync Trigger  198
   System  196
   video  198
      digitizer  33, 198
'Beacon Assist'  116, 119, 198
Beam sharpening  46, 103, 198
Beam width control  198
Bendix  189
Binary-to-octal code converter  125
Bistable multi-vibrator  183
BITE  91, 155, 158, 163, 198
Blip  198
Bloomer  198
Blooming  198
Bracket  198
   decoding  198
   pulses  198
Buffer amplifier  185
Built-in test equipment  91, 155, 198
BVD  199

C-pulses  199
Cable runs  139
Caboose pulse  199
Call  199
Capture  199
Carrier frequency  19
Cascode circuit  145, 146
Centre mark  199
   marking  199
Challenge  199
Channel, auxiliary  111, 112
   main  100, 111, 112

Circulating store  35
Circulator  100
Civil air traffic control  85, 127, 173, 191
Clock gate  150
Clock pulse generator  78, 150, 168, 182
   with delay line  183
Clock pulse stage, alternating  185
Coaxial cable  123, 183
Co-channel operation  52, 82
Code  21, 199
   acceptance gate  189
   changes  191
   coincidence gate  179
   garbling  59, 199, 201, 203
   response  199
   selector switch  120, 128, 157
   -selector unit  182
   -setting unit  116, 117, 120
   train  199
Coder  81, 96, 199
   test  156
   trigger  128
   unit  151
Coding-gate signals  190
Coding matrix  182
Coincidence  149
   circuit  150
   gate  149
   suppressor  150
Coincident display  199
Commercial airlines  173
Common-base circuit  145, 146
Common decoder  87, 89, 114, 121, 122, 127, 199
Common pivot mount  43, 44
Common system  200
Communication failure  200
Comparator  164
Complementary outputs  185
Components, minimum number of  152
Compression  184
Concentric resonator  153
Construction  173
Continuous interrogation  119
Control
   components  118
   pattern  200
   pulse  49, 200, 205, 211
   transmitter  83

Control—*cont.*
  unit  84, 118, 126, 131, 132, 167, 174, 189
  unit C4083/APX  157
Convertibility  166
Count down  200
Coupling loop  153
Coverage
  probability  164
  range of  113
Covering range  195
C/S  200

D-pulses  200
Data
  arrangement  133
  check  125
  -processing unit  89
  reliability  68, 117, 125
  transmission equipment  127
  transmission feature  200
Dead time  16, 65, 200
Dead-time generator  150
Dead-time pulse  153
Decimal converter  123
Decoder  83, 132, 200
  delay line  189
  losses  179
  matrix  117, 122
  single  87
  test  156
  tray  114
  unit  148
Decoding  149, 150
Decoupling  145, 153, 189
Deflection system  43
Defruiter  52, 67, 83, 96, 112, 196, 200
  criteria  55, 201
  trigger  97, 98
Defruiting  201
  efficiency  58, 201
Degarbling  88, 119
  circuit  114, 201
Delay line  54, 88, 97, 147, 148, 149, 183, 185
Demodulator  145
Department of Defence  138
Descent  194
Design  139

Detection level  105
Detector output  161
Dielectric strength  154
Difference
  diagram  45
  pattern  201
  signal  83
Digital
  decoding  167, 169, 175
  defruiter  112
  indication  125, 127
  indicator unit  84
  shift-register  167
Digitalization  168, 169, 175
Digitizer  201
Diode
  matrix  189
  technique  151
Diplexer  82, 103, 132, 133, 143
Directional
  antenna  36, 38, 51, 82
  characteristic  38
  diagram  40
Disc-seal triode  153, 181
Display unit  80, 84, 120
  control  81
Displayed beam/width  201
Ditch  201
Diversity transponder  165, 166
DME  13
DOD  138
Doppler effect  80
Double
  defruiting  201
  directional coupler  159
  -slash presentation  201
  stroke  123
Drip proofing  131
DT  201
DTF  201
Dual installation  174
Dynamic range  184
Dynamic shift-register  113, 185, 187
Duty cycle  140

Echo  65, 147
  suppression  147, 148, 177, 202, 208
  amplifier  66, 148
  circuit  66, 145

Effective beamwidth 202
Electrical shaft 80
Electromagnetic compatability 130
Electronically guided antenna 70
Emergency 202
    call 31, 202
        decoder 31, 118
        important types of 118
        repetition signal 116
        signal 89
Emergency, general 194
Emitter-follower 145
*En route* flight 194
Environment 202
Environmental conditions 85, 90, 91, 94, 131, 136, 163
Equipment, basic 96
    classification of 138
EUROCAE 138
European Organization for Civil Aviation Electronics 138
Excitation signal 92
Exploration 15
Explosion, resistance to 138
External interlock pulses 150
External suppression 202
Externally-controlled generator 182
Extractor 34, 44

FAA 12, 85
FAA aviation authorities 133, 134
Failure (communication) 200
Fall time 17, 19
False
    alarm 118
    code 202
    garbling 202
    moding 202
Falsified pulse train 125
Fan-shaped beam 15, 79
Feed-back conditions 153
Filter 82, 143, 188
Filtered display 202
Five-stage binary counter 182
Fixed objects 81
Flight 193
    check 193
    guard 183

lane 24
lanes 193
level 194
level FL 25
path 200
plan 193
type of 193
Flying altitude 125, 138
    method/code correlation 191
    procedural turn 193
Folded pillbox antennae 202
Four-post network 83
Framing pulse 22, 182, 198, 202
    coincidence 115, 116
    signal 88
    decoding 88, 198
    spacing 159
Framing pulses 202
Frequency
    adjustment 153
    discriminator 160
    doubler 180, 181
        stage 181
    shift beacon 13
    tolerance 180
    tripler 180
    spurious 145
Fruit 52, 202
    density 53, 203
    suppression 114
'Fruit' replies 203
Fundamental attenuation 61

Gain-time control 203
Garble sensing 203
Garbling 203
    suppressor circuit 116
    suppressor unit 116
GAT 203
'Gate' 154
Gate circuit 203
General aviation 173, 180, 188
    transponder 203
Ghost 203
Go-no-go test 199, 203
Gray code 27
Green indicator lamp 159
Grey zone 50

Grid/anode cavity  153
Grid-bias voltage  153
Grid-cathode circuit  153
Grid modulation  154, 183
GTC  81, 83, 203
    time function  112
    trigger  97, 113

Half-power point  41
Harmonic filter  103
Hazeltine Corporation  139, 163
Heterodyne frequency  145
Heterodyne oscillator  143
High tension voltage  156
Hit  203
Hits per scan  15, 63, 65, 69
Housing  139, 173, 179
HT generator  153
Humidity requirements  131, 136

ICAO  12
    Annex 10  133
    specification  147
    standard atmosphere  25, 27
IDENT  203
Identification  125, 195
    friend or foe  13, 203, 207
    individual  194
    information  14, 138
    of position  31, 204
    pulse  204, 210
    response  204
IF amplifier  145, 146, 181
    unit  144
IF post amplifier  104
IF signal  145
IFF  13, 191, 203
    transponder  139, 163
    transponder  AN/APX-46, characteristics of  140
    transponder AN/APX-90  163, 165
    transponder STR700  166
    Mark X-SIF System  204
    transponder AN/APX-46  139, 143, 165
    transponders, comparison of  165, 166
IFR  192
    flight  194
In-band beacon  13

Indicator lamp  118
Information
    circuit  187
    flow  164
    pulses  22, 204
Input resistance  146
Interference signal  123
Integral test and monitoring equipment  163
Integrated circuit (IC)  94, 151, 163
Integration, large scale  167
Interconnection leads  167
Interference blanker  204
Interlace  204
    ratio  204
Interleave  204, 213
Interrogation  14, 21, 132, 163, 199, 205
    encoding  21
    mode  21, 22, 81, 204
    path  204
    side-lobe suppression  47, 204
    pulse  81, 82
        pair  205
        repetition frequency  34, 68
    recurrence frequency  16, 205
    repetition frequency  16
Interrogator  205
    responsor  205
    unit  77
        type 1990  93
Instrument flight rules (IFR)  24, 192
Internal pressure  155
I/P  205
    and emergency signals, repetition of  156
    reply  153
    response  89
    switch  157
I/R  205
IRF  205
Isotropic radiator  36
ISLS  100, 205
    dead time  66

Jitter  113
Jitter-free  153, 169

Kill  205
Killer circuits  205

King Radio Corporation  187
KT 75  187

LEA  128
Lead time  96
Leading edge  18, 108, 109
LIA  127
Light gun  205
    pen  32, 84, 123, 125, 205
Line-equalizing amplifier  128
Line input amplifier  127
    transformation  145
Load fluctuations  155
Lobe formation  62
    switching  67
Location  14, 192
Logarithmation circuit  184
Logarithmic amplification characteristic  104
Logarithmic IF amplifier  104, 181
Logistics  85, 89
Longitudinal radiator array  40
Low-pass filter delay element  185
Low-pass filter unit  181
Low sensitivity  205

Magnetic store  55
Magnetron  78
Main beam killing  67, 205
Main control unit  118, 119, 120
Main-lobe interrogation  148
Maintenance  85, 90, 163
    of older units  163
    principles  94
Manual control  112
Mark X basic response  196
Marker  123
    gate  32
Mark X-IFF  206
Mark X-SIF  13, 119, 206
Matching  41
    factor $m$  42
    transformer  145
Meteorological correction values  89
Microphone button  204
Military air traffic control  94, 191
Minimum
    performance standard  134, 138
    separation  90, 193

triggering level  206
Ministry of Defence  138
Minor lobe  41
'Minus' tolerance  149
    limit  150
Mismatch  42
Mixer  143
    current indicator  156
    diode  145
MoA-Gilham code  28, 29, 125, 197, 206
Mode  199, 206
    coincidence signals  188
    control gate  177, 178
    gate  81, 118, 120, 123, 132, 150, 169, 178, 188, 206
    generator  128, 169, 178
    pulse  97
    indicator lamp  119
    interlace  56, 98, 206
        interrogation  113
    signal  128
    switch  125
    trigger signal  113
    1 coincidence  150
    1 gate generator  150
Modular sub-units  93
Modular system  93
Modulating pulse  154
Modulation  153
Modulator  78, 82, 153
Monitoring  90, 158, 159
    criteria  161
Monopulse
    antenna  50, 82, 198, 201
    decoder  206
    evaluation circuit  45, 112
    receiver  46
    technique  45, 83, 103, 111
Monostable multi-vibrator  177, 179, 185
MPS  134, 138
MTI (moving target indicator)  80
    radar unit  35
MTL  206
Multi-channel
    active/passive decoder  96
    decoder  121, 122
    receiver  112
Multiple
    cable  122

Multiple—*cont.*
  decoder 206
  transmission 68
Multiplex transmission 122
Multiplexing device 183
Multiplier 145
Mutual suppressor gate 178

Narco avionics 180, 184
National standards 134
Navigation 192
Navigational instruments 193
Non-common decoder 206
Non-synchronous reply-code overlap 59, 115, 207
'Normal' 123, 147
Normal pulse 169
Normal sensitivity 207
Numerical display 89, 196

Omni-directional antenna 36, 51, 82, 132, 133
Omni-directional characteristic 67
Omni-notch characteristic 207
Operating condition 158
Operating position 125
Operational reliability 90
Original pulse 169
Out-band beacon 13
Output circuit 146
Over interrogation 207
Overload 146
  control 153
    circuit 181
  suppressor 150

Parallel
  bit 118
  form 88, 116
  registers 175
  video signal 88, 116, 118, 125
  store 125
Parrot 207
Passive decoder 83, 121, 207
Passive decoding 31, 80, 89, 116, 119, 125
Phantom 207
Phased array 70
*p-i-n* diode 100, 101
  switch 103

Plan position indicator 207
'Plus' tolerance 149
  limit 150
Polar coordinates 207
Polarization 41, 43
Position report 193
Positional response 207
Positional signal 207
Power 159
  supply 94
PPI 81, 207
Precision approach radar (PAR) 26
Pressure, reduced 131
  altimeter system 132
  altitude 25
Pre-trigger 35, 81, 97, 198, 207
PRF 207
Primary radar 207
  unit 77, 207
Priority 98
  allocation 120
Privately-owned aircraft 134
Probability density 63
Probability grid 63
Processing time 35
Programming unit 98
Propagation conditions 195
Proportionality factor 110
Pulse analyser 168
Pulse
  distortion 128
  divider circuit 81
  pair 149
  pattern 116
  repetition frequency 207
  -shaping circuit 114
  tolerance limits 118
  top 108
  -width discriminator 114, 145, 147
$P_1$ 207
$P_2$ 207
$P_3$ 207

QNH correction 125
  value 26
Quantizing error 55
Quartz-crystal oscillator 99, 180
Quartz-crystal-stabilized transmitter 180

Radar
  clock pulse generator  81
  dead time  113
  frequency band  19
  monitoring, positive system  195
  observer  194
  operator  84, 122, 125
  pulse-repetition frequency  96
  receiver  80
  returns, enhancement  195
Radar (ASR)  196
Radio
  failure  194
  service  134
  Technical Commission for Aeronautics  134
  telephone  193
Range
  bin  207
  box  208
  correlation of  34, 82
  increment  54, 55, 113, 207
  marker  78
  measurement  16
  ring  81
Rapid read-out  208
Raw video  208
  indication  29, 119
RC delay element  185
RCA, Radio Corporation of America  173, 174
'Read-out' circuit  176
'Read-out' process  152
'Read pulse'  153
Receive subtraction  208
Receiver  83, 103, 132
  channel  164
  dynamics of  67
  gain time  81
  gate  81
  mixer  180
  sensitivity  118, 159
    evaluation of  159
    control  78
    reduction  148
  signal, minimum  145
  test  156
  units, specifications  84
Receiving path side-lobe suppression  208

Reception dead time  178
Recovery time  208
Redundancy arrangement  175
Reference level  47
Reflections  208
  at earth's surface  62
Reflector  43, 44
Rejection  150
  characteristic  56
  probability  208
  tolerance  169, 176
Reliability  90
Remote control equipment  96
Remote-controlled 'control'  127
Remoting equipment  126
Reply, emergency  153
  I/P  153
  normal  153
  code  208
    allotment of  193
    overlap  207
  hit  208
  hits per scan  208
  path  208
  -path side-lobe suppression (RSLS)  51, 208
  pulse train  116, 132, 189
Reporting points  193
Reserve
  channel  179
  gain  208
  stage  185
Resolving power  117
Resonance rise  143
  transformer  153
Resonator, /4  161
Responder  208, 211, *see also* Transmitter, Transponder
  unit  220
Response  15, 23, 132
  decoding  67
  encoding  22
  information, evaluation  29
  information, presentation  29
  probability  64
  pulse train  15
Responsor  208
  power switch  49
  switch  164

221

Responsor—cont.
  unit   143, 154
Ring around   47, 208
Rise time   17
Rolling ball   32
Rotating joint   79
Round reliability   209
  of system   60, 66
RSLS   100, 103, 111, 208, 209
  evaluation   112
RTCA   134, 138
Rules of procedure   193

Safety   90, 158
Saturation   105
Scanning   24, 63, 69
  programme   79
Screening effect   70
Second train   209
Secondary radar   209
  interrogator unit   93
  characteristics of type 1990   128
  process   194
  system   196
  transponder   132
    specifications of   133
  unit   77
Secondary surveillance radar   209
Select aircraft   209
Select code   209
Select mode   209
Selective identification feature   209
Selector switch   157
Self-oscillating power output stage   180, 183
Semiconductor components   94
Sensor   92
Serial form   88, 116
Serial transmissions   123, 125
Series parallel conversion   88, 114, 199
Setrin SLS   209
Setting up leads   118
Shadow   163
Shift
  pulse   169
    continuously operating   168
  register   152, 169, 179, 185
    capacity   169
    timing   169, 185

Shock resistance   131
Short-range radar   196
Shrimp boats   209
Side-lobe   41, 46, 209
  attenuation   41
  interrogation   148
  suppression   46, 132, 147, 177, 200, 208, 209, 211
SIF   210
  process   191
Signal, special   194
  echoes   147
  limiter   147
Single defruiting   210
Slant range   59, 80
Slash   210
Slave antenna   43
Slotted coupler   161
SLS   210
  antenna   83
  coincidence   150
  criterion   185
  dead time   177
  gate   150
  gate switch   177
  pulse   81
  switch   81
    control signal   81
  trigger   97
  unit   101
Space position   115
Special position   210
  identification pulse   199, 210
Specification R1212   85
Specification T50 C74a, characteristics of   135
Specifications   85, 134
Spherical emitter/radiator   36
SPI   23, 116, 194, 210
  pulse   29
Spike   145
  discriminator   210
  suppression   87, 145, 147, 177
Split   210
Spurious pulse   168
Spurious radiation   137
Squawk   210
  2, Squawk 3, Squawk Alpha, Code ...   210

flash 210
low 210
mayday 210
normal 210
Squitter 67, 210
SSR 12, 210
  2100 174
  ground unit 205
  secondary radar system, coding 21
  systems, possible development 69
  transponder 139, 173
  uses of 195
  video output 198
Standard ATR housing 173
Standard Elektrik Lorenz 112
Standby 210
Start information signal 113
Static register 185
STC 78, 80, 211
Stimulus 92
Storage capacity (defruiter) 113
Storage tube 55
  defruiter 211
Strangle 2, Strangle 3 211
Strangle flash 211
Sub-harmonic 16, 34, 81, 97
Successive detection 106
  principle of 104, 106, 181
Sum pattern 211
  signal 83
Summation diagram 45, 50
Summing resistance 107
Superheterodyne receiver 83, 99, 132
Suppression 208, 209
  signal 66
Suppressor gate 150
Surge 155
Surveillance radar 196, 209
Switch drive-circuit 103
Switching speed 101
Switching time 101
Synchronization 112
Synchronous filter 53, 83, 200
Synchronous reply-code overlap 60, 115, 211

Tacan 13
Tapping 116, 148, 149, 150

Target
  centre 199
  controlled interrogation 58
  display 81
  -echo point 198
  killing 211
  selector 123, 127
    instruction 125
  unit 123
Technical service order 136
Temperature-stabilized LC generator 150
TEST 159
Test and maintenance procedures 139
Test
  mode 140
  programme 92
  transponder 91
  signal generator 92
  unit 155
Testing 158
  principles of, 94
  and monitoring units 157, T29, 158, 161
Thick film technique 179
Three-pulse side-lobe suppression 211
Three-pulse technique 100
Threshold 145, 211
  amplifier 148
Time increment 55, 113
Timing 151
  frequency, minimum 185
  raster 150
Toggle switch 174
Tolerance 169
TR switch 79
Trailing edge 108, 110
Transistor technique 151
Transistor/diode technique 152
Transistorization 181
Transmission, azimuth 80
  channel 60
Transmitter 78, 82, 98, 153, 180
  oscillator 153
  output stage 181
  test 156
  unit 153
  units, specifications 84
Transponder 65, 203, 211
  reserve 174

223

Transponder—*cont.*
   dead-time  211
   delay  211
   evaluation  158
   inflight monitor lamp (TIM)  159
   reply limit  211
   specification  136
   suppression  98
Transverse radiator array  41
Trigger level  211
TSO  136
Two-channel passive decoder  114, 116
Two-channel receiver  165

UAT-1  179
Use, different categories of  136

Variable attenuator  146
Very-high-voltage surge  155
VFR  192

VFR flight  194
Vibration resistance  131
Video
   bandwidth  19
   map  80
   mixer  80
   signal  156
Viewing angle  63
Visual flight rules  192, 194
Voltage-standing-wave-ratio  42, 159

Waiting  194
Weight, reduction of  167
Wideband amplifier  128

X pulse  22

Zero point, temporal  78
Zero transition detector  179